Biodiversity, Access and Benefit-Sharing

T0228256

The Nagoya Protocol to the Convention on Biological Diversity (CBD) is rapidly receiving signatures and ratifications. Many countries are preparing to implement the protocol through national research permit systems and/or biodiversity laws. Yet there is still considerable confusion about how to implement the Protocol, regarding access and benefit-sharing (ABS) procedures, and minimal experience in many countries. This book seeks to remedy this gap in understanding by analysing a number of ABS case studies in light of the Nagoya Protocol.

The case studies are wide-ranging, with examples of plants and micro organisms for medicinal, cosmetic, biotech and food products, from or for development in, Australia, North Africa, Madagascar, Thailand, the United States, Switzerland, Panama and the Pacific. These will encourage countries to develop national systems which maximise their benefits (both monetary and non-monetary) towards conservation and support for local communities that hold traditional knowledge. In addition, the author analyses new expectations raised by the Nagoya Protocol, such as the encouragement of the development of community protocols by indigenous and local communities. As a result, stakeholders and policy-makers will be able to learn the steps involved in establishing ABS agreements, issues that arise between stakeholders, and the types of benefits that might be realistic.

Daniel F. Robinson is Senior Lecturer at the Institute of Environmental Studies, University of New South Wales, Australia, and Visiting Research Fellow, International Centre for Trade and Sustainable Development (ICTSD), Geneva, Switzerland. He is author of *Confronting Biopiracy* (Earthscan, 2010).

Biodiversity, Access and Benefit-Sharing

Global case studies

Daniel F. Robinson

Routledge
Taylor & Francis Group

LONDON AND NEW YORK

earthscan
from Routledge

First published 2015
by Routledge
2 Park Square, Milton Park, Abingdon, Oxon OX14 4RN

and by Routledge
711 Third Avenue, New York, NY 10017

First issued in paperback 2017

Routledge is an imprint of the Taylor & Francis Group, an informa business

British Library Cataloguing-in-Publication Data
A catalogue record for this book is available from the British Library

Library of Congress Cataloging-in-Publication Data
A catalog record has been requested for this book.

ISBN 13: 978–1–138–30452–9 (pbk)
ISBN 13: 978–0–415–71427–3 (hbk)

Typeset in Bembo
by Swales & Willis Ltd, Exeter, Devon, UK

Cover images, from top to bottom: Boat purchased for the community from
ICBG upfront compensation, Ambavarano village, Diana region, Madagascar
(Daniel Robinson 2013); argan fruit before being dried, cracked and pressed
for argan oil, near cooperative Taitmaine, Taroudant region, Morocco
(Daniel Robinson 2010); ecotourist trail and tower walk in the Falealupo
rainforest, Savii island, Samoa (Daniel Robinson 2012); roadside genetic
resources, central Madagascar near Antananarivo (Daniel Robinson 2013).

Contents

Illustrations

Figures

Tables

Boxes

Acknowledgements

I would like to thank the Commonwealth Department of Environment (DSEWPAC) and AusAID for their support on this research as case studies that were a component of their contribution to the GIZ-led multi-donor ABS Capacity Development Initiative. I thank, in particular, Ben Phillips and Mark Taylor for their support and for comments on the Pacific case studies and the Australian chapter.

I would like to thank the University of New South Wales (UNSW) and the Faculty of Science for providing early career and faculty small grant funding which helped bridge the funding gap to make this book possible. Especially, I thank the staff at the School of Biological, Earth and Environmental Sciences (BEES) and the Institute of Environmental Studies (IES) at UNSW for assisting with my daily work and for often providing the freedom and time to pursue this research. I thank in advance the ICTSD for allowing me to have sabbatical in Geneva in 2014, during which time I will no doubt be doing final edits and checks on this manuscript.

Madagascar: I have to extend an enormous bout of gratitude to the team at Missouri Botanical Garden (MBG) in Madagascar. This fieldwork would not have been at all possible without their assistance in the small amount of time that I had. As can be seen from that chapter, I met hundreds of people and gathered an enormous amount of information in Madagascar. This includes Chris Birkinshaw, Jeannie Raharimampionona, Jeremie Razafitsalama, Mamisoa Andrianjafy, Christian Camara and Germain Randrianandrasana from MBG. I also thank Luciano Andriamaro from Conservation International, Prof. Felicitee Rejo and Dr Radu from CNRE, and Dr Michel Ratsimbason at CNARP, among others. I thank Amir Antoy for his translations. I also thank Prof. David Kingston for commenting and providing much information about the benefits generated in this case. I also greatly appreciate the time given by the people whom I interviewed and their willingness to share their perspective on the projects. If I could name them all, I would.

Thailand: I want to thank several staff at Biotec (Dr Kanyawim Kirtikara, Sronkanok Tangjaijit and Dr Bubpha Techapattaraporn) and Shiseido (Mr Koichi Yoshino) for their time and clarifications. I also want to thank my colleagues including Buntoon Srethasirote, Witun Lianchamroen and Somchai

Rattaneskul for their comments and updates on the laws in Thailand (and Jakkrit Kuanpoth indirectly for his collaboration, support and advice over the years).

Samoan Mamala case: I thank Clark Peteru (SPREP) for providing information, contacts and background and for assisting with the logistics of fieldwork in Samoa. I thank Kolone Va'ai for also assisting with fieldwork logistics and providing his knowledge of the case. Paul Cox gave me useful information, details of the agreements and comments on my case study. I thank the Matai and people of Falealupo for their warm welcome and assistance.

Moroccan Argan case: The following people all assisted immensely with information about the case, as well as assistance in the field: Prof. Zoubida Charrouf (University of Rabat), Eric D'efrenne (Yamana), Latifa Anaouche (Targanine), Charlotte D'Erceville-Dumond (BASF), Raquel Ark (formerly at Cognis) and Rachel Barre (L'Oreal). I thank the women in the cooperatives who spoke with us and shared a little insight into how their lives have changed. I thank the Union for Ethical BioTrade (UEBT) for inviting me to present at their 'Beauty of Sourcing with Respect' conference, which was the genesis of this chapter.

Australian ABS system: I thank Ben Phillips (DSEWPAC), Geoff Burton (UNU), Libby Evans-Illidge (AIMS) and Richard Lipscombe (Proteomics) either for the review of the chapter, or useful information relating to the laws and examples for the chapter.

Vanuatu: I acknowledge Donna Kalfatak and Trinison Tari from the Department of Environment and Conservation who assisted with contacts, information and advice for this case study. I thank several others who assisted the fieldwork, for the interviewees and communities who gave their time. I thank Phillipe Bouchet (French NMNH) for sharing a copy of their agreement and for his comments.

ICBG PNG: I acknowledge Prof. Louis Barrows, Prof. Lohi Matainaho and Dr Eric Kwa who enthusiastically and openly shared their project with me and explained the procedures and national ABS reform in PNG.

ICBG Panama: I acknowledge Profs. Bill Gerwick, Thomas Kursar and Phyllis Coley who provided useful updates and comments regarding their project. For general information and comment on the ICBGs, I thank Joshua Rosenthal and Flora Katz.

I thank Prof. Graham Dutfield and Prof. Michael Blakeney for taking the time to provide useful comments on drafts of my chapters.

I also thank Geoff Burton and Dr Andreas Drews who have given me comments and advice relevant to some of the case studies and the book. They also expanded my knowledge of ABS and the Nagoya Protocol during interesting conversations, workshops, field trips and dinners in the Pacific. I also thank my colleagues in Natural Justice: Johanna Von Braun, Harry Jonas, Kabir Bavikatte and Holly Jonas for their passion, inspiration and interest in this field.

I thank Tim Hardwick and the team at Routledge for their assistance, editing and enthusiasm to publish my work.

Finally, I thank my wonderful family: my wife Rachel and my boys Miles and Felix, for their support, patience and love.

Daniel Robinson

Acronyms and abbreviations

ABS	access and benefit-sharing
AIMS	Australian Institute of Marine Science
ARA	AIDS Research Alliance
Biotec	National Center for Genetic Engineering and Biotechnology, Thailand
Bonn Guidelines	Bonn Guidelines on Access to Genetic Resources and Fair and Equitable Sharing of the Benefits Arising out of their Utilization
CBD	Convention on Biological Diversity
CGIAR	Consultative Group on International Agricultural Research
CI	Conservation International
CITES	Convention on International Trade in Endangered Species of Wild Fauna and Flora
CNARP	Madagascan Centre Nationale d'Application et des Recherches Pharmaceutiques
CNRE	Madagascan Centre Nationale de Recherches sur l'Environment
CNRO	Madagascan Centre Nationale de Recherches Oceanographiques
CSR	corporate social responsibility
DAS	Dow AgroSciences
disclosure requirement	A proposed patent requirement that applicants disclose the source and/or country/place of origin or legal provenance of genetic resources used in an invention
DoA	Department of Agriculture
DoE	Australian Commonwealth Department of Environment
DSEWPAC	Australian Commonwealth Department of Environment, Sustainability, Water, Populations and Communities
EC/EU	European Commission/European Union
EEZ	exclusive economic zone
EIG	economic interest group
EPBC Act	Environmental Protection and Biodiversity Conservation Act
EPO	European Patent Office
FAO	Food and Agriculture Organization
FIC	Fogarty International Centre
FDA	US Food and Drug Administration

FRI	Forest Research Institute
GATT	General Agreement on Tariffs and Trade
GC-MS	gas chromatograph – mass spectrometer
GI	geographical indication
GIZ	Deutsche Gesellschaft für Internationale Zusammenarbeit (German implementing agency for international development)
HPLC	high performance liquid chromatograph
HTS	high throughput screening
IARC	International Agricultural Research Centre
ICBG	International Cooperative Biodiversity Groups
ICNP	Intergovernmental Committees of the Nagoya Protocol
ICTSD	International Centre for Trade and Sustainable Development, Geneva
IGC	World Intellectual Property Organization Intergovernmental Committee on Intellectual Property and Genetic Resources, Traditional Knowledge and Folklore
ILCs	indigenous and local communities
INDICASAT	Panama's Institute of Advanced Scientific Research and High Technology Services
IP/IPR	intellectual property/intellectual property right
IRD	Institute for Research for Development
ITPGRFA	International Treaty on Plant Genetic Resources for Food and Agriculture
IUCN	International Union for the Conservation of Nature
MAT	mutually agreed terms
MBG	Missouri Botanical Gardens
MTA	material transfer agreement
Nagoya Protocol	Nagoya Protocol on Access to Genetic Resources and the Fair and Equitable Sharing of Benefits Arising from their Utilization, to the Convention on Biological Diversity
NCI	National Cancer Institute
NIH	US National Institutes of Health
NGO	non-governmental organization
NMNH	National Museum of Natural History, France
NSTDA	National Science and Technology Development Agency
NT BRA	Northern Territory Biological Resources Act
ONE	Madagascan Office National pour l' Environnement (ONE)
PIC	prior informed consent
PVP	Plant Variety Protection
R&D	research and development
SAGE	Service d'Appui a la Gestion de l'Environnemont
SIMM	Shanghai Institute of Materia Medica
STRI	Smithsonian Tropical Research Institute
TEEB	The Economics of Ecosystems and Biodiversity
TK	traditional knowledge

TRIPS	Trade-Related Aspects of Intellectual Property Rights
UEBT	Union for Ethical Biotrade
UNARGEN	Unidad Nacional de Recursos Genéticos, Panama
UNCLOS	United Nations Convention on the Law of the Sea
UNDRIP	United Nations Declaration on the Rights of Indigenous Peoples
UNESCO	United Nations Educational, Scientific and Cultural Organization
UPNG	University of Papua New Guinea
USAID	US Agency for International Development
USDOA	US Department of Agriculture
WHO	World Health Organization
WIPO	World Intellectual Property Organization
WTO	World Trade Organization

Part I

Access and benefit-sharing

1 Introduction, context, methods and trends

This book has been written as a follow-up to my original book *Confronting Biopiracy: Challenges, Cases and International Debates* published in 2010. While my original book received some generous praise from reviewers, it is likely to leave many readers feeling that there is little hope for the conduct of 'fair and equitable' biodiscovery access and benefit-sharing. This is by virtue of its focus on the polemical concept of biopiracy. My own sense of the negative was reiterated in the comments of a book review in the journal *Conservation Biology*. As Davis (2011, p. 643) described of *Confronting Biopiracy*:

> This is not a book on how to prevent biopiracy, and it does not explore the perspective of potential users who may be flummoxed by the issues and obstacles. I hungered for some success stories.

Although this had not strictly been my intention, this is quite an apt comment on the first book. With this in mind I have here set out to find some success stories and to assist with the issues and obstacles of the 'flummoxed' potential users of genetic resources and associated traditional knowledge (TK). Therefore the intent of this book is to outline the trends and activities of the 'users', the principles and obligations that are being or will be imposed by the Nagoya Protocol, to analyse several existing biodiscovery access and benefit-sharing case studies and then to identify lessons that can be drawn from the case studies. Before proceeding to explain the structure of the book and the methods and approaches taken, it is worth briefly explaining biodiscovery, access and benefit-sharing.

What is biodiscovery access and benefit-sharing?

The Convention on Biological Diversity (CBD) came into force in 1993 and has 3 main objectives. The objectives are the conservation of biological diversity, the sustainable use of its components and the fair and equitable sharing of benefits arising out of the utilisation of genetic resources. While the first two objectives are likely to be plainly understood to the ordinary person, the third is a compromise that arose because of demands from mainly biodiverse and 'mega-diverse' countries (mostly located in the tropics) from the 'global South' or developing world. The collection of biological samples of plants, animals and microbes by researchers often (but not always) from the 'global North' and advanced economies has a modern history[1] starting with the 'green revolution' and leading up to the 1990s as commercial interest expanded in pharmaceutical companies. This sort of collection activity for research and development (R&D) purposes is commonly described as 'bioprospecting' and more recently as 'biodiscovery'. It was perceived that this R&D could yield commercial products worth significant sums of money – millions or billions of dollars.

Genetic resources can be understood as genetic materials from plants, animals or microbes that have a potential value to be used. They form the basis for many products and services that humans utilise and benefit from. Although scientific research has now made us increasingly aware of the economic and conservation value of genetic resources, humans have arguably had a strong understanding of the function and usefulness of plants and animals for centuries or even millennia. Genetic resources may provide for the natural metabolism and production of biochemical compounds that can be utilised for a range of purposes: cosmetics and skin care, pharmaceuticals, industrial enzymes, crop varieties, food additives and products, poisons and pest control, among others. Genetic resources also represent components of ecosystems which contribute to many beneficial ecosystem services such as waste decomposition, carbon sequestration and nutrient cycling.

Through numerous advancements in the life sciences as well as developments in areas of study such as ecological and environmental economics, the *potential* attribution of economic value to genetic resources has expanded significantly. It was this perceived potential that led to the 'grand bargain' in the CBD text – a compromise between facilitating and/or allowing access to genetic resources in return for benefits (both monetary and non-monetary) that might in turn contribute to the conservation of biodiversity (including genetic resources). Benefit-sharing from biodiscovery was seen as a potential 'win-win' because biodiscovery activities are generally considered to be non-exhaustive extractions of the genetic resource. In other words, only small samples might be needed to conduct the R&D. If the R&D subsequently leads to commercially valuable and viable products, the CBD encourages 'sustainable use' of the resource, which might mean sustainable farming of a plant, breeding of an animal or the laboratory propagation of microbial cultures. Thus, the potential for harm to the environment from over-cultivation of a genetic resource is perceived to be low.

The CBD text specifies that access to genetic resources for the purposes of R&D should be sought with the permission of the provider country (prior informed consent (PIC)). Access permission can be granted subject to a benefit-sharing agreement made under mutually agreed terms (MAT) between the user (the researcher) and the provider country. This process has come to be known simply as 'access and benefit-sharing' (ABS) to negotiators, biodiscovery researchers and relevant policymakers. However, the original CBD text left various aspects of ABS ambiguous and open to interpretation, such as the following:

- The scope of application of ABS to R&D on 'derivatives', namely bio-chemical extracts, from genetic resources;
- The need to obtain PIC from local providers of genetic resources, as well as competent national authorities; and
- The need to also seek permission and share benefits with providers of TK which may be associated with the genetic resources used for R&D, among others.

This has then led to subsequent negotiations to clarify the terms of ABS, and in 2010 the Parties to the CBD agreed upon the *Nagoya Protocol on Access to Genetic Resources and the Fair and Equitable Sharing of Benefits Arising from their Utilization*. The achievement of the Nagoya Protocol to the CBD was another key driver for this book, as it creates a new challenge for CBD Parties that are ratifying and implementing the Nagoya Protocol. They will need to put in place policies and/or laws and administrative measures so that ABS works in practice. Understanding the practical aspects of ABS is critical to those involved in the administration of ABS policy and regulations, as well as for those who conduct R&D on natural products who will have to work with new regulatory arrangements and expectations. Case studies of recent ABS-type agreements and/or activities provide an excellent tool for reflecting upon the challenges that occur when seeking to achieve 'fair and equitable' biodiscovery R&D.

Structure of the book

The book is split into 3 main parts. The first part includes introductory content exploring ABS in Chapter 1 and in Chapter 2. The remainder of Chapter 2 outlines the approaches and methods utilised in the research and writing of the book, followed by a section engaging with more conceptual and critical literature and finally a section which explores ABS concepts and trends. Chapter 2 discusses the Nagoya Protocol including a review of recent literature, analysis of the new scope for ABS and some potential challenges for its implementation.

The second part includes several biodiscovery ABS case study chapters. These cases have been selected from around the world so as to demonstrate the existence of different R&D activities, different national approaches to ABS and different ABS-type agreements. Case study 'provider' countries that were analysed include Madagascar, Papua New Guinea, Morocco, Vanuatu, Samoa, Thailand and Panama. Other brief examples are also discussed. In most, but not all, cases fieldwork and interviews were conducted in the provider country and user country stakeholders were also interviewed. For each case study, I review the following:

- The aims and intent of the biodiscovery projects,
- Their relevance as an ABS-type project,
- Their adherence to relevant laws,
- The access procedures followed,

- Aspects of the PIC of local people, and mutually agreed terms,
- The benefit-sharing agreement and how it aligns with the Annexe of the Nagoya Protocol, and
- The impacts or potential impacts of the monetary and non-monetary benefits for local conservation, research and communities.

The third part then includes a discussion chapter which reflects on lessons that can be drawn from the case studies, comparing and contrasting different aspects of them. Several factors or themes are used to explore the issues including the following:

- Permits and PIC,
- Enabling versus protective national systems,
- Widely held traditional knowledge,
- Secret, sacred and locally held traditional knowledge,
- Mutually agreed terms, power relations and 'fairness and equity',
- Access and upfront benefit-sharing,
- Utilisation and benefit-sharing,
- Monitoring and checkpoints,
- Compliance, and
- Intellectual property.

Part 3 follows with some final conclusions on future issues and challenges.

Approach and methods

Research and fieldwork towards this book were undertaken largely in 2012 and 2013. Secondary information about the biodiscovery activities and ABS agreements from available sources was collected. Relevant local media reports were particularly sought after to determine whether there were any positive or negative local perceptions of a project and its impacts. In most cases fieldwork was then also conducted in the 'provider' country to confirm and validate aspects of a benefit-sharing agreement and its impacts. Fieldwork typically involved semi-structured interviewing of officials from government agencies which regulate access and may have acted as providers. In addition, interviews were conducted with researchers, local government, local providers, in some cases 'traditional knowledge-holders', project partners including companies, relevant NGOs, beneficiaries of the R&D or agreements and others. Ethical approval was obtained for the research through my institution – UNSW. Either a written or a verbal PIC was obtained for each interview, depending upon the level of formality and criteria such as whether the informant agreed to be identified and quoted. Direct observations were made at specific sites that demonstrated tangible benefits as a result of an ABS-type agreement, and this has been documented with photography throughout the book.

To the best of my abilities I have engaged with a diverse group of relevant stakeholders and also searched relevant local media for potential issues or commentaries on biodiscovery projects. I have also regularly cross-checked information with the biodiscovery researchers and other local stakeholders to ensure they are able to confirm facts or refute claims that may have been made. The research had several limitations – key stakeholders were identified and interviewed during relatively rapid fieldwork activities for each case study, often in remote and difficult to reach locations. However, not all stakeholders could always be interviewed, nor all field sites examined and photographed. Fieldwork was sometimes undertaken in locations that were facilitated by those involved in the projects, raising a question of influence or bias. Interviews were often conducted with the help of a translator, especially in local communities, having the effect that meanings are sometimes misconstrued during translations. Despite these constraints, great care was taken to minimise the potential for bias, to seek information from a wide group of stakeholders (often including non-beneficiaries) and to collect relevant information independent of those with potentially vested interests in a project.

The research sought to determine alignment with CBD principles. Depending upon the stage that the biodiscovery activity was conducted, the CBD, Bonn Guidelines and also the Nagoya Protocol are all considered in terms of alignment. While this is inevitably a subjective process (e.g. what is 'fair and equitable?'), there are certain principles and indicators which we can assess – access permit and PIC (including local PIC if required), MAT, monetary and non-monetary benefits (as per the Annex of the Bonn Guidelines and Nagoya Protocol) and core CBD objectives.

Critiques and conceptualisation of ABS

My last book, *Confronting Biopiracy*, and also this book, have been written using relatively normative language intended to be accessible to the many stakeholders involved or interested in biodiscovery and ABS. However, I do want to spend a small space here adopting a slightly more critical conceptual lens and language.

The conceptual framework of political ecology underpins much of the thinking in this and my earlier book, being concerned with the political economy and 'socio-natures' of the ABS process (i.e. not just analysis of productive access to and use of natural resources and commodities), and decision-making and political influence at different scales (adapted from Watts and Peet, 2004; Bakker, 2010). Although an expansive set of literature, much recent political ecological work might be characterised as describing or locating:

> . . . land management and environmental degradation (or sustainability) in terms of how political economy shapes the ability to manage resources (through forms of access and control, through forms of exclusion, and through forms of exploitation), and through the lens of cognition (one person's accumulation is another person's degradation).
>
> (Watts, 2003, p. 592)

This book follows this trajectory towards an understanding of how 'ABS' has been developed, utilised, sought and provided, as a market-based tool, a biodiversity-oriented discourse and as a potential mechanism towards conservation, livelihoods and social justice. After reviewing each of the case studies, I will return to this idea in the conclusions of the book. For now, how might those from political ecology and similar disciplinary approaches view the concept of ABS?

To provide some historical context, it must be recognised that there has been an extended period of colonial and post-colonial exploitation of natural resources, people and their knowledge, from populations and countries of the 'Global South'. The book *Confronting Biopiracy* starts by documenting the colonial spice trade, plant hunting and the brutal oppression of the people and colonies that the European powers conquered (Robinson, 2010, pp. 1–10). Not only were there deliberate efforts to control both populations and territories for the extraction of natural resources, including biological resources such as spices and ornamental plants, but there were other violent oppressions such as slavery and war, and unintended consequences such as diseases and epidemics (see Coates, 2005; Rodney, 1972). Colonists and missionaries had their own direct and destructive impacts on systems of government, societal structures, cultural norms and practices. The post-independence period – for most countries after World War II – also saw a host of neo-colonialisms: the erosion of peasant systems of agriculture (many already disrupted and shifted during the colonial period) through the green revolution (Shiva, 1991), the establishment and globalisation of market-oriented economies through the GATT and then World Trade Organization (WTO) (Braithwaite and Drahos, 2000) and the encouragement of spiralling debt levels in developing and least-developed countries leading to the imposition of debilitating structural adjustment policies of the International Monetary Fund (Beder, 1996; Bradshaw and Huang, 1991; Naiman and Watkins, 1999). These are relevant here because they both established the 'development' context and have influenced the degradation of the environment in the Global South. They also importantly provide the frame through which many actors discuss the perpetuation of inequities and injustices which appear in the debates surrounding ABS. For example, indigenous authors continue to argue that neo-colonialism is perpetuated through the way research objectifies indigenous people, pursues insensitive or extractive research practices and activities and writes histories and knowledge that exclude other people (Smith, 1999; Said, 1978). Critiques have also highlighted the inherent neo-colonialism in the way international environmental laws and governance mechanisms are established to the exclusion or exploitation of indigenous and local people, and in doing so, the way 'global knowledge' is constructed as if all or most parties accede to such worldviews (Harry, 2011; see also Monfreda, 2010).

Another concern that has been raised regarding cultural concerns surrounding what is researched and then in some cases patented or monopolised. There are indeed some people within specific cultural groups who have strong spiritual or customary connections to aspects of nature including plants and animals. The ability to modify – especially forms of genetic modification – may be perceived as an affront to spiritual beliefs surrounding a life-form for some people. Similarly, the ability to patent and claim ownership rights over the reproducible components of a plant variety, or even an extract from a plant that has certain cultural or spiritual importance, can be seen as being in conflict with the beliefs of some groups. This can result in a sense of personal injury to those individuals who 'embody' the sense of wrongdoing (discussed in Robinson et al., 2014). These changes have accelerated quickly in our lifetimes, through rapid developments in biotechnologies and through rapid global changes to intellectual property (IP) laws, affecting states that sign up to those laws (including many developing and least developed countries).

If we turn to questions surrounding environmental degradation and inequalities, there are some that would take the relatively benign view that, despite this history and geography of exploitation, environmental reforms through advances in science and technology will enable the compatible pursuit of economic growth and environmental conservation. This reflects an 'ecological modernisation' perspective very much in line with the concept and rhetoric surrounding 'sustainable development' and is often perpetuated from an array of actors in the privileged North. Problematically, this allows close to a 'business as usual' approach to consumption and production, with profitability and the aims of economic and political elites taking precedence over social and environmental considerations (Agyeman and Carmin, 2011). This is despite overwhelming evidence about the seriousness of global environmental impacts. For example, the Global Biodiversity Outlook 3 report presents a particularly bleak picture of biodiversity loss: none of the 21 sub-targets accompanying the overall target of significantly reducing the rate of biodiversity loss by 2010 have been definitively achieved globally (CBD Secretariat, 2010). Relevant to ABS a specific example of the reduction in crop genetic diversity can be found in China, where the number of local rice varieties being cultivated has declined from 46,000 in the 1950s to slightly more than 1,000 in 2006 (CBD Secretariat, 2010).

To further conceptualise the above, the colonial and post-colonial historical geography described aligns with a Marxist view of the primitive accumulation of capital. But rather than being just a historical period, more contemporary interpretations of Marx reinterpret the concept of primitive accumulation as being more than just a process of accumulation by expanded production or by colonial dispossessions, but as an ongoing process that facilitates further dispossessions towards capital accumulation by a wealthy elite (Harvey, 2003; Corson and MacDonald, 2012). This is often broadly described as a 'neoliberalisation of nature' (described below) and is often described as a form of neocolonialism. Harvey (2003) explained that market liberalisation ultimately produces chronic crises of over-accumulation: elites holding accumulated capital need to find new ways of investing it to generate further surplus. Thus it requires the continual release of new assets to seize and to convert to profit – new commodities, markets and opportunities for speculative investment have been identified with nature, in this case, biodiversity. This shift is posited as 'the only option' by environmental economists, most notably in recent years through The Economics of Ecosystems and Biodiversity (TEEB) project. The problem, as perpetuated by the authors of TEEB, is not with the capitalist system or with markets as a basis for exchange and distributions, even though it is clear that many of our environmental problems are caused and perpetuated by human overconsumption (Dauvergne, 2008) as well as expansive population growth. Rather it is that nature has not been adequately priced and is 'economically invisible' (Pavan Sukhdev, cited by Corson and MacDonald, 2012, p. 169). Thus it is not surprising that projects like TEEB are being thought of as instruments for capital expansion and 'agents of nature's restructuring', underpinning what Büscher (2011) called 'one of the biggest contradictions of our times': the idea that nature can be conserved by increasing capital circulation (Corson and MacDonald, 2012, p. 181). For many the market system *is the problem*, and indeed the number of 'assumptions' that environmental economists list for their proposals to actually work would leave many with serious doubts about their potential overall effectiveness (see Bishop, 2012, pp. 5–7). For example, the idea of 'substitution' of ecosystems, biodiversity and ecosystem services is always going to be polemical from an ecosystems dynamics standpoint – what unknown synergistic effects and benefits are we losing through the destruction of an ecosystem and 'replacement' with another (ten Brink, 2011)? Further, there remain concerns from various indigenous and local communities (ILCs), NGOs, conservationists and academic spheres about the need to also recognise and value the intrinsic, cultural and spiritual values of biodiversity – not just to allocate prices to it. Clearly there will be those who view the pricing of nature as essential for its conservation and others who continue to criticise and defy it. ABS forms one part of this larger valuation process – for those who champion it, it is seen as a potential 'win-win' for conservation of biodiversity (and/or sustainable use), as well as for important R&D. Increasingly it is being seen as a potential 'win' for local provider communities – but only in some cases and only given the right circumstances.

Following political ecology and related literatures, it is possible to see why the concept of bioprospecting and ABS might be described as being part of a concerted project towards the neoliberalisation of nature – indeed it has been described as 'green developmentalism' following a 'postneoliberal environment economic paradigm' (McAfee, 1999, p. 154). On providing a political economy understanding of the CBD, Boisvert and Vivien (2012) noted that it was the combination of alarming biodiversity erosion, harbingers of market opportunities and the expectation of profits drawn from genetic resources that gave rise to a demand for policy. They noted that economic discourse at the time stressed the systematic undervaluation of diverse biological resources, and that specific international agreements were needed to deal with the market and policy failures – in their parlance, biodiversity represents a 'common' or 'open access good' (non-exclusive) and which receives rivalry among economic agents and users. Certain aspects of the set of issues and potential policies for dealing with biodiversity erosion appealed to many economists: 'the emergence of a relatively new technique (genetic engineering), turning some elements of the environment (the genes) into resources, creating new rents, making institutional arrangements obsolete and requiring a redefinition of property rights (including intellectual property rights)' (Boisvert and Vivien, 2012, p. 1165).

The CBD was negotiated around the same time as the Uruguay Round of Trade Negotiations which ultimately led to the establishment of the WTO in 1994. Although starkly different forums, the WTO Agreement on Trade-Related Aspects of Intellectual Property Rights (TRIPS) was significant for biodiscovery R&D, and in turn for ABS. Following R&D, IP (typically through patents or plant variety protections) provides a legal mechanism for companies and researchers to assert exclusive monopoly rights over new 'inventions' or plant varieties. This is perceived to be the primary vehicle for the commodification of nature and green developmentalism in the ABS context (McAfee, 2003; Hayden, 2003a). The CBD explicitly recognises that patents are likely to be sought and encourages technology transfer (including technologies under patent) to the countries that provide genetic resources (CBD, Article 16.3).

To explain what some of the critics of ABS and commentators mean when they discuss 'neoliberalisation of nature' as a general trend, Castree (2008, pp. 142–143) provided the following characterisation:

- *Privatisation* (that is, the assignment of clear property rights to social or environmental phenomena that were previously state-owned, unowned, or communally owned. Now owners of hitherto unprivatised phenomena can potentially come from *anywhere* across the globe).
- *Marketisation* (that is, the assignment of prices to phenomena that were previously shielded from market exchange or for various reasons unpriced. These prices are set by markets that are potentially global in scale, which is why neoliberalism is often equated with geographically unbounded 'free trade').
- *Deregulation* (that is, the 'rollback' of state 'interference' in numerous areas of social and environmental life so that (i) state regulation is 'light touch' and (ii) more and more actors become self-governing within centrally pre-scribed frameworks and rules).
- *Reregulation* (that is, the deployment of state policies to facilitate privatisation and marketisation of ever-wider spheres of social and environmental life).
- *Market proxies in the residual public sector* (that is, the state-led attempt to run remaining public services along private sector lines as 'efficient' and 'competitive' businesses).
- *The construction of flanking mechanisms in civil society* (that is, the state-led encouragement of civil society groups (charities, NGOs, 'communities', etc) to provide services that interventionist states did, or could potentially, provide for citizens; these civil society groups are also seen as being able to offer compensatory mechanisms that can tackle any problems citizens suffer as a result of the previous five things listed).

If we consider the political economy context in which the ABS aspects of the CBD were established and are being implemented nationally against these typologies of neoliberalisation, there is considerable alignment. However, this did not happen solely in or through the CBD fora – arguably much of the change in deregulation, marketisation and reregulation occurred through the GATT, TRIPS and other rules established under the WTO agreement, among other processes. Yet some of the results in the CBD text are quite evident, starting with the annihilation of any concept that plant genetic resources were the 'common heritage of [hu]mankind', which was fostered in the late 1970s and early 1980s by discussions towards the non-legally binding International Undertaking on Plant Genetic Resources for Food and Agriculture. The CBD instead recognises state sovereignty over biological resources – a 'pragmatic' way to seek clearer definition of property rights (physical and intellectual) and thus access. Discursively, the language of the CBD does generally facilitate a 'molecular-genetic reductionism' through its encouragement of 'facilitated access' and regular references to the 'utilisation of genetic resources' – 'resources' being a way of thinking that strips a material from its spatial and social context (McAfee, 2003; McAfee, 1999; Whatmore, 2002). This fulfils the 'marketisation' aspect of a neoliberal typology of the regulation of biodiversity. Biotechnologies and biochemical sciences also provide possibilities to commodify and commercialise resources at the molecular scale that previously may only have been commodified as seeds, foods, basic plant extracts (such as oils), farm produce and as parts of plant and animal species and their distinctly bred varieties. As Parry noted:

> it is now possible to extract genetic or biochemical information from living organisms, to process it by replicating, modifying, or transforming it, and to produce from it minor modifications of this information that are themselves able to be utilized as raw materials, commodified as resources.
>
> (Parry, 2000, p. 383)

The CBD text focuses on the utilisation of genetic resources for biotechnological (and other) purposes and encourages countries to *facilitate* access. The Nagoya Protocol now clarifies utilisation of genetic resources to include derivatives, and within that definition, 'biochemical extracts' (discussed further in Chapter 2). Thus there is a discursive broadening of 'natural' content for which R&D is encouraged, seeming to further perpetuate the scope of green developmentalism that might occur. On the other hand, this might be seen as a way of 'catching' a broader range of benefits for developing countries for extractions that have already been occurring for decades.

It could be debated whether the ABS provisions of PIC (which allows 'providers' of genetic resources to say no or to place conditions on research access), MAT (which ensures that the parties come to a mutually beneficial agreement) and benefit-sharing (which can be monetary and/or non-monetary as determined by the parties) are in fact a neoliberal de/reregulation. It does facilitate a marketisation of genetic resources. Yet, the introduction of these norms into the research access process seeks to promote greater 'fairness and equity' and this is at least a reregulation against the potential for outright unauthorised exploitation of a genetic resource (and associated TK). This is where ABS certainly starts to feel like an '*incomplete* neoliberalisation' of nature – fairness and equity are not typically regulated or found in open markets where competition rules.

Despite this, it has been noted that inequitable bargaining power can be a major problem in ABS negotiations (e.g. Greene, 2004). Increasingly, governments and local communities have been receiving NGOs and donor-funded projects that provide capacity building and awareness about ABS to reduce these potential imbalances of power. This again aligns with the characterisation above with respect to 'flanking mechanisms in civil society' – except that the Nagoya Protocol will potentially require some substantive regulatory, compliance and 'checkpoint' impositions on states. This again suggests that the negotiation process on an international regime on ABS and text of the Nagoya Protocol reflects an '*incomplete* neoliberalisation of nature', which we will return to in Chapter 2.

While this book acknowledges these neoliberal processes, the arguments put forward here are cautiously put regarding the existence of any 'unilateral' or even concerted neoliberal project that has resulted in the ABS component or related articles of the CBD text, and the subsequent Nagoya Protocol. There is certainly more evidence of direct industry influence on the establishment of the TRIPS Agreement (see Braithwaite and Drahos, 2000). Rather, it is suggested the CBD, and its creation, implementation and enforcement, is one of many laws and regulations like other 'contemporary forms of rule [which] are inevitably composite, plural and multi-form' (Larner, 2000, p. 20). As Hayden (2003a, p. 49) noted, 'bioprospecting is one of myriad efforts on behalf of a wide range of actors to valorize nature within firmly economic renderings of accounts.' Regarding the ABS aspects of the CBD, Hayden (2003a, p. 63) further explained that it 'effectively literalizes the market-mediated vision of biodiversity . . . as it banks on biotechnology and intellectual property as key engines for valorising biodiversity – and thus as indispensible in promoting conservation and non-destructive, sustainable development.' While this mindset has been present in the negotiations towards the CBD text, there have also been voices seeking social and environmental justice, which is at least partially (if weakly) recognised in the ABS language on 'fairness and equity' in the original CBD text. On these sets of more divergent views of biodiversity, McAfee (1999) acknowledged that the CBD has become:

> . . . a gathering ground for transnational coalitions of indigenous, peasant, and NGO opponents of 'biopiracy' and the patenting of living things, and advocates of international environmental justice. They have begun to put forward counterdiscourses and alternative practices to those of green developmentalism.

There is a considerable back-history to what ultimately was agreed as the ABS provisions of the CBD (see Reid et al., 1993; Eisner, 1990; Robinson, 2010) and it has been described as the 'grand bargain' (Jeffery, 2004) because the end result was a fairly ambiguous and loaded text that encourages R&D and commercial use of genetic resources (as pushed by various industry sectors, lobbyists and mostly Northern governments), but requires access permissions and benefit-sharing as a trade-off. The recognition of national sovereignty over biological resources in the CBD text is also to the potential benefit of governments of much of the biodiverse South. The text makes a gesture towards the respect and maintenance of 'traditional knowledge' of ILCs in Article 8(j) that is full of caveats, but arguably, has been used to leverage further indigenous knowledge 'protections' in other/later fora (e.g. UNDRIP, WIPO and the Nagoya Protocol).

Here in this book I draw similar lines to Larner's (2000, p. 20) analyses of neoliberal processes and power that the CBD and its terms relating to ABS can be thought of as 'a consequence of the contestation between dominant and oppositional claims, rather than being simply imposed from above'. The fact that the United States has not ratified the CBD points to the fact that the requirements for benefit-sharing have been controversial for some governments of the North and certainly for industries located there. Though many have noted that the United States has been present as an observer and its lobbying can be seen through a Foucauldian lens as influencing or 'governing at a distance' (Dean, 1999), the recent achievement of the Nagoya Protocol also reflects considerable positioning, counter-lobbying and resistance by developing countries, biodiverse countries, indigenous organisations and various NGO groups (Bavikatte and Robinson, 2011). For example, considerable lobbying by indigenous organisations such as the International Indigenous Forum on Biodiversity, and NGOs such as Natural Justice among others, led to inclusions in Article 12 of the Nagoya Protocol for the respect of indigenous and local community customary laws and encouragement of development of community protocols relating to TK. This element of progress for indigenous rights capitalises on existing incremental recognitions in international law such as the United Nations Declaration on the Rights of Indigenous Peoples (UNDRIP) and will be legally binding once the Nagoya Protocol is in force. The substantial significance of this achievement has been lost on many activists and scholars due to the technical nature of the Nagoya Protocol, yet this binding legal text might be seen as a lever through which further rights can be assigned to ILCs in other negotiations.

These relatively ground-breaking achievements in the text suggest there has been a perceptible shift in the actors involved in ABS, and we have increasingly seen the voices of indigenous people and local communities involved in individual negotiations and at the various CBD meetings relevant to ABS. Groups of these actors see the potential for tangible local 'benefits', both monetary and non-monetary, received by indigenous people and local communities, and towards conservation or 'sustainable use' of biodiversity. ABS agreements are being actively sought by some individuals, communities and governments, and resisted by others. There has been and will be complex results. Indeed as researchers come forward with a new mandate to share benefits upon/subsequent to research access, there can be an 'incitement to form groups with the purpose of negotiating' with them, as Hayden (2007, p. 746) noted. Through negotiation with the accessing parties (researchers and companies), we may see 'communities' formed where they may not have clearly existed such that future beneficiaries can be 'named' even though royalty payments, compensation and other benefits have often failed to be materialised for ABS recipients. Questions of 'fairness and equity' can then arise as a result of imbalances or inequities that arise in these newly formed and potentially disparately made-up communities (see Greene, 2004). Collectivisation, as Hayden (2007, 2003b) described it, and engagement of these groups is occurring and will occur as industry and researchers gradually 'buy-in' to the idea of ABS and/or are forced into it by binding national ABS laws and research permit requirements. There may be some mixed consequences as a result of this (we will return to this in Chapters 11 and 12). It may be that those people merged into a collective arrangement for the sake of potential benefit-sharing and those excluded from it may come to odds due to the transcribing of new boundaries (who benefits and who does not?) and the creation of new expectations (when do benefits occur and how large are they?) (e.g. see Strathern, 1999 on compensation and IP in Papua New Guinea). However, the broadening of existing indigenous rights now recognised under the Nagoya Protocol may see a shift in the balance of power in these transactions – time will tell.

Although the 'biopiracy' discourse (characterised in Robinson, 2010) may have generated much angst for researchers and industry, it has arguably facilitated a transition towards more CBD aligned R&D activities and/or some in industry have moved away from natural products research (e.g. pharmaceuticals). We now also see the rapid collection of signatures and ratifications of the Nagoya Protocol, and development of subsequent national ABS laws, and research permit systems. Transparency tools established by the Nagoya Protocol such as the Clearing House Mechanism (CHM) should make it relatively easy to identify and query non-compliance with ABS conditions once the regime is in place (although it will not work retrospectively and 'temporal scope' has been contentious given the quantity of materials already transferred to ex situ genebanks and private collections). Most rational researchers and industry will inevitably be seeking to avoid the public relations issues of having a project identified as non-compliant under the CHM and then run the risk of having their research called 'biopiracy'. This might ultimately mean less blatant and less common cases of biopiracy.

Although all of the cases explored in this book have occurred before the Nagoya Protocol entered into force, and some even prior to the CBD, we can still examine them as ABS-type agreements or arrangements and draw lessons from them. Specifically, we can hope to see through some of the cases that benefits have been delivered which go some way towards the (recognisably partial, loaded and sometimes ambiguous) elements of social and environmental justice that have been sought through the negotiated texts on ABS. However, it is also unreasonable to expect that any agreement will ever be perfect – common questions around fairness and equity will almost inevitably be raised in all biodiscovery ABS examples, and this process of ABS is likely to continue to be a steep learning curve for researchers, industry, government agencies and also local communities. Gradually, as the Nagoya Protocol is adopted and awareness grows among researchers and industry, it seems likely that we will see fewer severe biopiracy-type critiques and more articles lauding ABS agreements that benefit conservation, local communities and society more broadly.

ABS contexts and trends

Since approximately the time that the CBD was developed in the early 1990s, there has been a declining interest in natural product's drug discovery as a source of novel compounds for the pharmaceutical industry – the industry originally thought to be the most likely candidate to provide significant monetary benefits under ABS. The reasons for this are at least two-fold: scientific advances and changes in technology have led to new screening processes for drugs and the ABS aspects of the CBD were perceived as somewhat of a regulatory barrier or complication to biodiscovery activities. On new scientific advances, Laird and Wynberg (2012, p. 4) explained:

> Genomics, which refers to the study of the totality of an individual's genetic makeup (genome), and the related fields of proteomics (the study of proteins), metabolomics (the study of metabolites – the substances produced by chemical reactions in the cells of organisms), transcriptomics (the study of the process of transferring genetic information from DNA to RNA) and phenomics (the study of phenotypes in relation to genomics) have emerged from the convergence of new molecular techniques, bioinformatics and automated laboratory tools for generating molecular data.

These new fields of research have altered the way drug discovery is conducted, and also the way research is conducted in other industry sectors such as cosmetics, agricultural biotechnology, plant breeding, crop protection, industrial enzymes, foods, beverages and nutraceuticals.

This has led to a decline in field-based collection activities, especially for pharmaceuticals. In other industry sectors such as cosmetics, there is still strong interest and demand for natural products, especially plants and botanicals. However, these are often obtained as raw extracts and if limited or no R&D is conducted on the extracts from genetic resources, then this arguably falls outside of the scope of ABS as defined by the Nagoya Protocol. Increasingly, the cosmetics industry is conducting extensive R&D on genetic resource (typically plant) extracts, and will have to comply with ABS provisions of countries that implement the Nagoya Protocol. In other sectors, some are suggesting that new technologies may mean greater interest in genetic resources and continued bioprospecting. For example, new DNA sequencing technologies are able to investigate millions of polymorphisms (variations within the same species) and dramatically expand our ability to select desirable attributes (Laird and Wynberg, 2012). This means more rapid selection processes in crop breeding and should benefit agriculture through the development of drought, flood or pest-resistant plant varieties from farmer's varieties or crop wild relatives.

Many in these sectors also note that much of the preliminary research is often conducted in universities. Once a lead is of sufficient interest, the research is then often licensed to or purchased by a company. This essentially means that public sector institutions have been shouldering the costs and risks, while industry has become more risk averse regarding natural products discovery.

In addition, it has been noted by many observers that extensive ex situ collections of genebanks and herbariums now exist from field collections that have been occurring for decades or even centuries. Extensive agricultural collections during the green revolutions have ended up in the International Agricultural Research Centre genebanks, now under the control of the Consultative Group on International Agricultural Research. These are regulated by the terms of the Standard Material Transfer Agreement and Multilateral System under the International Treaty on Plant Genetic Resources for Food and Agriculture (ITPGRFA – discussed further in Chapter 2). However, there are many government and private company collections of a range of genetic resources (not just agricultural) where the access regulations are not always as transparent. These may act as a source for continued access for natural products development well into the future, removing the need to negotiate with provider countries and communities who regulate in situ collections (Beattie et al., 2005).

To understand this further, here we briefly explore some of the trends in bioprospecting and natural products R&D by industry sector.

Pharmaceuticals

The drug discovery process involves many phases including target identification, lead identification, small-molecule optimisation and pre-clinical through clinical development before seeking marketing approvals (Burbaum and Tobal, 2002). This is a lengthy process that can take an average of 14 years of R&D, and an estimated $800 million to $1.7 billion to achieve a drug (Cragg et al., 2012). A commonly cited figure in the pharmaceutical industry is that from 10,000 or more 'hits' tested in early drug discovery, only a handful may lead to a drug that reaches the market. To elaborate, the phases of drug discovery are briefly explained below:

- Target identification usually involves identification of relevant proteins in a person's body which are associated with a disease, or proteins in a micro-organism causing a disease. Cellular networks of proteins are usually analysed and pathways identified where relevant to protein function relevant to the mechanisms of a disease (Burbaum and Tobal, 2002);
- High throughput screening (HTS) is the most common drug discovery method in the pharmaceutical industry, in which tens of thousands of compounds a week, using enzyme or receptor-based assays are designed to

uncover compounds with specific mechanisms of action (Kingston, 2011, p. 498). A 'lead' or 'hit' is identified when chemical compounds are found to bind to the identified target and show potential for modifying the disease;

- The 'small molecule' with potential is then optimised in terms of molecular structure to ensure that it has the right pharmacokinetic properties for appropriate safe and effective absorption and metabolism in humans;
- Pre-clinical trials involve tightly regulated testing of the identified drug candidate, usually on animals, for safety and efficacy;
- Clinical trials involve phase 1 and 2 testing on small groups of candidates that are willing to participate in the trials, including both candidates with the target disease and those that do not. This is followed by phase 3 trials with larger target groups of people, to further refine safety and efficacy and to identify side-effects;
- Following this, if trials are successful, regulatory and marketing approvals are typically sought from relevant government agencies, such as the US Food and Drug Administration (US FDA) before release of the drug to market.

Natural products have historically played an important role in this drug discovery sequence. Despite a drop-off of interest, especially by industry in the 1990s and early 2000s, there is still a group of interested scientists, most based in universities. As many large companies have closed down their natural products programmes (e.g. Merck, GlaxoSmithKline, AstraZeneca, Eli Lilly and Pfizer), there are still others such as Novartis that are active. Some, such as Merck reputedly still retain partnerships focusing on natural products, and even Pfizer lists an interest in partnering for next generation natural product screening technologies at the time of writing,[2] though Pfizer appears to have closed Wyeth's natural products division when it was acquired.

Newman and Cragg (2012) pointed to the fact that natural products have provided a source of novel structures, but not necessarily the final drug entity, for many new drugs. They note that from around the 1940s to date, of the 175 small molecules (in cancer research), 131, or 74.8 per cent are other than synthetic, with 85 or 48.6 per cent, actually being either natural products or directly derived therefrom (2012, p. 311). In their categorisation of sources they include:

- Biologicals (usually a large peptide or protein, for example, isolated from an organism/cell);
- Natural products;
- Natural product botanicals;
- Derived from a natural product – usually a semi-synthetic modification;
- Totally synthetic drug, often found by random screening or modification of an existing agent);
- A drug made by total synthesis, but the pharmacophore was from a natural product;

- A subcategory of 'natural product mimics' relating to the above two categories – where there is a competitive displacement of a natural substrate by a synthetic compound; and
- Vaccines.

According to their breakdown, only 387 of 1,355 (29 per cent) of all new approved drugs in the 30 years from 1981 to the end of 2010 are totally synthetic. Despite some suggestion that the natural product mimic categories are an exaggeration of the role of natural products, there is still strong evidence that natural products are providing leads in some form to the majority of drugs. Even if the 'natural mimic' categories are dropped, there are still 700 (almost 52 per cent) of drugs in this timeframe from natural sources (Newman and Cragg, 2012, p. 312).

The decline in natural products-based drug discovery in terms of funding from major granting agencies and also the support of this research by major pharmaceutical companies has occurred for several reasons, as adapted from Kingston (2011, pp. 498–499):

- Crude natural product extracts are complex mixtures containing hundreds of compounds which may require additional steps and time to make them compatible with HTS, which has become the dominant mode of screening because it can screen tens of thousands of compounds (typically derived from combinatorial chemistry) per week.
- Resources have been diverted to combinatorial chemistry, which is essentially computer-aided generation of a large number of synthetic compounds to be tested for bioactivity – usually by HTS. Many individual generic molecular structures can be input and the outcome of each is dozens or hundreds of slightly modified derivatives, making up libraries of compounds to be screened.
- Technical challenges in the isolation of bioactive molecules from highly variable source materials (e.g. plant leaves or roots, soils containing microbes, marine sponges and symbiotic microbes); including the risk that the active compound eventually isolated is already a known compound.
- Resupply problems can be an issue when a potential drug is identified, because much larger quantities of the material may be required for pre-clinical development and then clinical trials. As noted above, the source material can be highly variable and obtained from remote locations. Finding additional source material can be costly and time-consuming. If the genetic resource is from a rare species, this can be highly challenging as this raises long-term conservation and sustainability concerns. Synthesis of the lead compound can sometimes resolve this problem.
- Regulatory or policy constraints (and also related public relations and ethical concerns) have arguably expanded since the conception of ABS under the CBD. For field prospection in national parks, certain marine areas, and foreign countries, a long process of negotiation for permissions, export of samples and benefit-sharing may be required.

- Financial pressures have led to massive cuts to R&D in industry (and also to a lesser extent affecting academia), and this has heavily impacted the relatively slow and time-consuming process of natural products lead identification and drug discovery.

Despite the industry-led shift away from natural products, many university-based scientists are developing new ways of fast-tracking natural products R&D and are also identifying new areas of interest from natural products. For example, Kingston (2011) suggested that alongside de-tanninisation of crude extracts, the creation of 'peak libraries' where crude extracts are prefractionated into a series of pure or almost pure compounds allows it then to be screened using HTS. He also suggested that natural products can be used as the scaffolds for combinatorial libraries, while in reality these have typically been based on computer-aided natural product-like scaffolds. Li and Vederas (2009) described approaches for the screening of micro-organisms, suggesting massive increases in the number of fermentations so as to identify new microbials which may be masked by common antibiotics found in soil, such as streptomycin. They also noted that while plants remain a major source of new drugs, cyanobacteria and marine organisms demonstrate interesting potential in terms of neurotoxic and cytotoxic compounds (Li and Vederas, 2009). Von Nussbaum et al. (2006) described in detail the potential for reversed genomics to provide an efficient generation of lead structures in antibiotics. Finally, Weber and Fussenegger (2009) highlighted that synthetic biology can provide cost-effective microbial production processes for complex natural products where there have been global drug supply shortages. For example, they noted the production of plant secondary metabolites from the antimalarial drug artemisinin (from the sweet wormwood *Artemisia annua* L), and the chemotherapeutic taxol derived from the pacific yew tree (*Taxus brevifolia* Nutt.); can be cost-effectively produced through biosynthetic means in common microbes (Weber and Fussenegger, 2009). These technological advances are likely to mean that natural products will have some continued role in drug discovery, even if this role is reduced by combinatorial chemistry and the HTS analysis of natural product-like synthetic libraries.

Agriculture

Within the broadly defined agricultural industries, there are identifiable areas of R&D in the production of seeds, for crop protection, agricultural chemicals and plant biotechnologies, among others. These industries have seen increasing global consolidation through mergers and acquisitions in recent decades. As the CBD Secretariat (2008, p. 15) noted, just 'ten companies control 55% of the commercial seed market and 64% of the patented seed market . . .' with 'the value of the overall commercial seed market in 2006 estimated at $30 billion.' As of mid-2006, the agbiotech industry was dominated by Monsanto, Syngenta, Bayer and DuPont, controlling much of the world seed market and plant breeding industry. As Murphy (2007, p. 174) noted, these companies are especially dominant in agbiotech IP rights, 'where they owned over 77% of all US utility patents in 2005'. In commercial agbiotech (transgenic crops), this trend is even more concentrated with Monsanto owning 90 per cent of the seed used in transgenic crops in 2005 (Murphy, 2007).

Although these agricultural industry sectors are clearly relevant under the ABS framework of the CBD, it is important to note that many of the key crops commonly commercially grown and on which much R&D is conducted, regulation of access and utilisation comes under the framework of the ITPGRFA of the UN Food and Agriculture Organization (FAO). Furthermore, industry sourcing of these types of plant genetic resources has often been from international and national genebanks or from private collections. This means that germplasm has likely been transferred between breeders, genebanks, research institutions and companies using material transfer and acquisition agreements (MTAs) rather than under ABS-type agreements.

Presently, only about 30 crops provide 95 per cent of human food energy needs, 4 of which (rice, wheat, maize and potato) are responsible for more than 60 per cent of our energy intake (FAO, 2014). Despite the dominance of certain crops and the likelihood of germplasm for breeding coming from genebanks, there is and will continue to be plant breeding undertaken using germplasm from other sources and these are likely to be regulated under national systems established in compliance with the Nagoya Protocol.

While R&D on genetic resources in these sectors has been highly varied, some trends can be observed. Agricultural biotechnology industry research budgets are estimated as some $1.95 billion annually, yet only a small proportion of this involved the harvest of seeds from the wild (Beattie et al., 2005). This research instead tends to focus on improving existing transgenic crops. Murphy (2007) suggested that R&D utilising transgenesis has been declining in many cases because genetically complex characters such as yield and fungal resistance (highly desirable for farmers) are still proving very difficult to manipulate by transgenesis, whereas modern breeding techniques such as DNA marker-assisted breeding has been more successful. Transgenic crops are often modified so as to portray herbicide tolerance and insect resistance, or in some cases to introduce essential vitamins and minerals for humans. They are typically sold as packages with herbicides and fertilisers that the same company produces. While the transgenic crop may have some advantages, from a farmer's perspective, there may also be risks (e.g. negative public perception of transgenic crops) and some perceived advantages of using non-transgenic crops with their genetically complex characters and resistances (Murphy, 2007).

Using a number of modern techniques (e.g. tissue culture technologies, chemical methods of analysis and selection of crop traits, DNA marker-assisted selection and molecular breeding techniques), commercial breeders are able to rapidly and cost effectively develop new varieties selected for specific beneficial traits and characteristics. Because of these new breeding technologies, commercial interest has continued to focus and even to further concentrate on those 4 major crops mentioned above, and to a lesser extent the other 30. This means re-examination of available germplasm, and declining interest in exotic genetic resources. Overall, as a result these companies will likely continue to cross their transgenic lines with existing available germplasm to improve the dominant herbicide or insect-resistant products that can be sold as a profit-maximising proprietary suite of products. However, they will also be mindful of some competitive advantages of applying modern breeding techniques to modern varieties (often protected by companies under plant breeders rights but obtainable subject to licence fees), sometimes to generic varieties and landraces (domesticated crops but often with considerable genetic variation), and with only some rare interest in wild germplasm (undomesticated plant varieties) for specific disease resistances.

The CBD Secretariat (2008) noted that the crop protection sector has relatively more interest in wild genetic resources than other areas of agricultural research, and thus, is of higher relevance in the ABS context. These might include specific fungus or insect resistances bred into plant varieties, or also new 'natural' sources of fungicide or pesticide to be sprayed onto plants. Pyrethrins, derived from a Chrysanthemum variety, are the classic example, and have become a commonly used household insecticide, as well as being used on various crops and in horticulture. The CBD Secretariat (2008, p. 114; and see McDougall, 2010) cited that R&D in 10 leading crop protection companies indicates an overall R&D expenditure of $2250 million, equivalent to 7.5 per cent of sales for these companies.

Cosmetics, skincare and related industries

Although it is hard to define the 'personal care' sector, it is largely made up of cosmetics, skincare, hair-care, fragrances and other personal hygiene products. These industries use wild harvested or cultivated biological resources in a wide variety of products, including cosmetics, feminine hygiene, hair products, baby care, nail care, oral hygiene, deodorants, skincare creams and gels and fragrances (Beattie et al., 2005). According to industry estimates, sales of 'natural personal care products' have exceeded $7 billion in recent years (Dayan and Kromidas, 2011). While this industry sector is under-researched, secretive and saturated with public relations 'spin', there are a few general trends that can be estimated. A number of relatively large brands that sell 'natural' and 'organic' cosmetics, skincare and personal care products have emerged in recent decades. Some of the largest natural product-related brands such as Burt's Bees (owned by Clorox), Aveda (Estee Lauder), Aveeno (Johnson and Johnson) and The Body Shop (L'Oreal) have estimated sales in the order of $150–350 million dollars per year (Duber-Smith, 2011). The acquisition of The Body Shop by L'Oreal, the largest cosmetics company in the world, and a company which has continually branded itself as both high-end and scientific in its approach, suggests the perceived continued market demand for branded natural and organic personal care products. Every year the Union for Ethical Biotrade (UEBT) conducts a survey on perceptions of cosmetics and biodiversity and produces an annual biodiversity barometer. Interestingly, 84 per cent of surveyed consumers say that they buy beauty products that contain natural ingredients, with interest especially high in China (98 per cent) and Brazil (95 per cent) (UEBT, 2013).

It is very common for these sectors to utilise large quantities of natural products as basic extracts or commodities for their products. Bishop (2012) gave the example of cosmetics companies such as Estee Lauder and L'Oreal using a succulent plant listed in CITES Appendix II (*Euphorbia antisyphilitica*) as an essential raw ingredient in a wide array of cosmetics (especially lipsticks), inks, dyes, adhesives, coatings, emulsions, polishes and gum base. While these industries have often extracted raw materials from nature in relatively large quantities, there has been a question regarding the relevance of ABS for cosmetics/personal care since the inception of the concept in the CBD. Are they 'utilising *genetic* resources' for R&D? Under the ambiguous definition provided by the CBD in 1992, many companies would argue that they are not – rather they are utilising raw natural materials and biochemical extracts or mixtures (not genes) for basic commercial application, or in some cases after some R&D (usually in the case of major companies like those noted above). However, with a clarification of this definition under the Nagoya Protocol, R&D which utilises 'biochemical extracts' now is included within the scope of 'utilisation of genetic resources'. This means that many cosmetic companies will need to operate within the scope of the Protocol and establish ABS agreements with the providers of these products, where R&D is being conducted.

UEBT has also produced a number of interesting papers on global patent trends in the cosmetics sector. The patent search identified a total of 2,101 full species names (predominantly plants) in the titles, abstracts and claims of 3,523 patent documents (UEBT, 2010). Many of these patent documents make claims to quite basic mixtures of raw ingredients, while others for sophisticated R&D claims to identify the bioactive molecules responsible for certain beneficial cosmetic or personal care applications. For example *Aloe vera* is noted in 386 patents (UEBT, 2010), but the majority of these are likely to be claiming basic mixtures that utilise the plant. The paper also notes that the country of origin can often be more than one country – indeed plant species can often be found in several countries, whole regions or even across different continents. Therefore, it is not always a straightforward transaction from one location to another, whereby a benefit-sharing agreement can be easily put in place. For example, the Shea tree (*Vitellaria* paradoxa), commonly used in the skin care and cosmetic industry for its moisturising qualities and essential fatty acids, is endemic to 19 countries in Central and West Africa.

Other sectors

There are several other significantly large and relevant sectors to consider. These include botanical medicines, nutraceuticals, industrial enzymes and bioremediation.

Botanical medicines are typically made from raw plant materials. Revenues from these products can be large, for example, annual sales of medicinal gingko, garlic, evening primrose, and Echinacea in Europe average $350 million, and global sales of raw botanical material by US suppliers amounts to approximately $1.4 billion (Beattie et al., 2005; ten Kate and Laird, 1999). Their relevance for ABS can be questioned on the basis that often no R&D is conducted, except perhaps for some basic safety and efficacy testing. However, there is potential for R&D in this market, and indeed it is not uncommon to see patent filings for herbal treatments – implying and claiming 'innovation'. There is potential for companies to examine the biochemical makeup of these raw materials and market more advanced products provided they meet safety approvals. What we have seen in recent years are also herbal medicines booms as a result of R&D towards pharmaceuticals from specific plant extracts. For example, hoodia (*Hoodia gorgoni*) and devil's claw (*Harpagophytum* species) from Southern Africa have both been investigated by researchers and patented. Subsequently, a surge in opportunistic interest in both plants as herbal medicines has occurred, and they are being over-collected from the wild.

Related to this is the 'nutraceuticals' industry which sells food ingredients or products which are claimed to provide health or medical benefits. The functional food and nutraceutical industry is a $75.5 billion industry (Basu et al., 2007). Again, this industry largely utilises and supplies basic plant extracts; however, there are many patents and considerable R&D evident relating to the specific biochemical activity of the extracts and supplements they provide.

Sources of novel microbes for the treatment of contaminated soil and water are another area of bioprospecting interest. Bioremediation is used to clean up contaminated industrial sites from sectors such as mining, petroleum extraction and supply, and other heavy industry. Beattie et al. (2005) gave the example of white rot fungi which is now of major commercial importance because they degrade many highly toxic classes of pollutants such as PCBs, dioxins and furans. Demain and Vaishnav (2009, p. 297) also described the significance of microbes for the industrial enzyme market:

> The total market for industrial enzymes reached $2 billion in 2000 and has risen to $2.5 billion today. The leading enzyme is protease which accounts for 57% of the market. Others include amylase, glucoamylase, xylose isomerase, lactase, lipase, cellulase, pullulanase and xylanase. The food and feed industries are the largest customers for industrial enzymes. Over half of the industrial enzymes are made by yeasts and molds, with bacteria producing about 30%. Animals provide 8% and plants 4%. Enzymes also play a key role in catalyzing reactions which lead to the microbial formation of antibiotics and other secondary metabolites.

Given their sourcing from biological sources, and the considerable R&D that can be involved in determining new uses for these enzymes, ABS is clearly of potential relevance.

The following chapter provides an analysis and literature review of the Nagoya Protocol to the CBD, including discussion of the relevance of different aspects and challenges for these sectors. This is then followed by the case studies of bio-prospecting ABS.

Notes

1   It has a much longer colonial history arguably starting with the spice trade and various botanical expeditions led by European powers to the 'New World', Asia, Africa and other newly 'discovered' regions. While the colonial collections traded spices, animals and ornamental plants commercially, and also conducted scientific research on them, the intent of the research was arguably less commercially oriented and on a smaller scale than we have seen in the last century (see Robinson, 2010, Chapter 1; and Dutfield, 2003).
2   Specifically see the Pfizer website: 'Small Molecule Therapeutics Discovery'. Available at: http://www.pfizer.com/partnering/areas_of_interest/small_molecule_therapeutics_discovery.jsp, accessed 5/7/2013.

References

Agyeman, J. and Carmin, J. (2011) 'Introduction: Environmental injustice beyond borders'. In Carmin, J. and Agyeman, J. (eds) *Environmental Inequalities Beyond Borders: Local Perspectives on Global Injustices*. MIT Press, Cambridge, MA, pp. 1–16.

Bakker, K. (2010) 'Debating green neoliberalism: The limits of "neoliberal natures"'. *Progress in Human Geography*, 34(6), 715–735.

Basu, S.K., Thomas, J.E. and Acharya, S.N. (2007) 'Prospects for growth in global nutraceutical and functional food markets: A Canadian perspective'. *Australian Journal of Basic and Applied Sciences*, 1(4), 637–649.

Bavikatte, K. and Robinson, D. (2011) 'Towards a people's history of the law: Biocultural jurisprudence and the Nagoya Protocol on access and benefit sharing'. *Law, Environment and Development Journal*, 7(1), 35.

Beattie, A.J., Barthlott, W., Elisabetsky, E., Farrel, R., Teck Kheng, C. and Prance, I. (2005) 'New products and industries from biodiversity'. In Hassan, R., Scholes R., and Ash, N. (eds) *Ecosystems and Human Well-being: Current State and Trends*, Vol. 1. Island Press, Washington DC, pp. 271–295.

Beder, S. (1996) *The Nature of Sustainable Development*, 2nd edn. Scribe, Newham Australia.

Bishop, J. (ed.) (2012) *The Economics of Ecosystems and Biodiversity in Business and Enterprise*. Earthscan/Routledge, London.

Boisvert, V. and Vivien, F.D. (2012) 'Towards a political economy approach to the Convention on Biological Diversity'. *Cambridge Journal of Economics*, 136(5), 1163–1179.

Bradshaw, Y.W. and Huang, J. (1991) 'Intensifying global dependency: Foreign debt, structural adjustment, and Third World underdevelopment'. *The Sociological Quarterly*, 32(3), 321–342.

Braithwaite, J. and Drahos, P. (2000). *Global Business Regulation*. Cambridge University Press, Cambridge.

Burbaum, J. and Tobal, G.M. (2002) 'Proteomics in drug discovery'. *Current Opinion in Chemical Biology*, 6(4), 427–433.

Büscher, B. (2011). 'The neoliberalisation of nature in Africa'. In Dietz, T., Havnevik, K. and Kaag, M. (eds) *New Topographies of Power? Africa Negotiating an Emerging Multi-Polar World*. Brill, Leiden, pp. 84–109.

Castree, N. (2008) 'Neoliberalising nature: The logics of deregulation and reregulation'. *Environment and planning A*, 40(1), 131.

CBD Secretariat (Laird, S. and Wynberg, R.). (2008) *Access and Benefit-Sharing in Practice: Trends in Partnerships Across Sectors*. CBD Secretariat, Montreal, Technical Series No. 38.

Convention on Biological Diversity (CBD) Secretariat. (2010) *Global Biodiversity Outlook 3*. CBD Secretariat, Montréal.

Coates, K.S. (2005) *A Global History of Indigenous Peoples*. Palmgrave Macmillan, New York.

Corson, C. and MacDonald, K.I. (2012) 'Enclosing the global commons: The convention on biological diversity and green grabbing'. *Journal of Peasant Studies*, 39(2), 263–283.

Dauvergne, P. (2008) *The Shadows of Consumption: Consequences for the Global Environment*. MIT press, Cambridge, MA.

Davis, K. (2011) 'Toward profits without piracy'. *Conservation Biology*, 25(3), 642–644.

Dayan, N. and Kromidas, L. (eds). (2011) *Formulating, Packaging, and Marketing of Natural Cosmetic Products*. Wiley, New Jersey.

Dean, M. (1999) *Governmentality: Power and Rule in Modern Society*. SAGE, London.

Demain, A.L. and Vaishnav, P. (2009) 'Production of recombinant proteins by microbes and higher organisms'. *Biotechnology Advances*, 27(3), 297–306.

Duber-Smith, D.C. (2011) 'The natural persona care market'. In Dayan, N. and Kromidas, L. (eds) *Formulating, Packaging, and Marketing of Natural Cosmetic Products*. Wiley, New Jersey.

Dutfield, G. (2003) *Intellectual Property and the Life Science Industries: A Twentieth Century History*. Ashgate, Aldershot, UK.

Eisner, T. (1990) 'Prospecting for nature's chemical riches'. *Issues in Science & Technology*, 6(1), 31–34.

FAO. (2014) '*Plants. Use Them or Lose Them*'. Available at http://www.fao.org/nr/cgrfa/cthemes/plants/en/, accessed 24/04/2014.

Greene, S. (2004) 'Indigenous people incorporated? Culture as politics, culture as property in pharmaceutical bioprospecting'. *Current Anthropology*, 45(2), 211–237.

Harry, D. (2011) 'Biocolonialism and indigenous knowledge in United Nations discourse'. *Griffith Law Review*, 20(3), 702–727.

Harvey, D. (2003) *The New Imperialism*. Oxford University Press, Oxford.

Hayden, C. (2003a) *When Nature Goes Public: The Making and Unmaking of Bioprospecting in Mexico*. Princeton University Press, Princeton NJ.

Hayden, C. (2003b) 'From market to market: Bioprospecting's idioms of inclusion'. *American Ethnologist*, 30(3), 1–13.

Hayden, C. (2007) 'Taking as giving. Bioscience, exchange, and the politics of benefit-sharing'. *Social Studies of Science*, 37(5), 729–758.

Jeffery, M.I. (2005) 'Intellectual property rights and biodiversity conservation: Reconciling incompatibilities between the TRIPS agreement and the convention on biological diversity'. In Ong, B. (ed.) *Intellectual Property and Biological Resources*. Marshall Cavendish Academic, Singapore.

Kingston, D.G.I. (2011) 'Modern natural products drug discovery and its relevance to biodiversity conservation'. *Journal of Natural Products*, 74(3), 496–511.

Laird, S. and Wynberg, R. (2012) *Bioscience at a Crossroads: Implementing the Nagoya Protocol on Access and Benefit-Sharing in a Time of Scientific, Technological and Industry Change*. The Secretariat of the Convention on Biological Diversity, Montreal.

Larner, W. (2000) 'Neoliberalism: Policy, ideology, governmentality'. *Studies in Political Economy*, 63(1), 5–25.

Li, J.W.H. and Vederas, J.C. (2009) 'Drug discovery and natural products: End of an era or endless frontier?' *Science*, 325(5937), 161–165.

McAfee, K. (1999) 'Selling nature to save it? Biodiversity and green developmentalism'. *Environment and Planning D: Society and Space*, 17(2), 133–154.

McAfee, K. (2003) 'Neoliberalism on the molecular scale: Economic and genetic reductionism in biotechnology battles'. *Geoforum*, 34(2), 203–219.

McDougall, P. (2010) '*The Cost of New Agrochemical Product Discovery, Development and Registration in 1995, 2000 and 2005–8*'. Final Report available at: http://www.iasis.ie/cms/uploads/34newsCost%20of%20R%20%26%20D%20for%20Pesticides%202011.pdf, accessed 24/4/2014.

Monfreda, C. (2010) 'Setting the stage for new global knowledge: Science, economics, and indigenous knowledge in "The Economics of Ecosystems and Biodiversity" at the Fourth World Conservation Congress'. *Conservation & Society*, 8(4), 276–285.

Murphy, D. (2007) *Plant Breeding and Biotechnology*. Cambridge University Press, Cambridge.

Naiman, R. and Watkins, N. (1999) *A Survey of the Impacts of IMF Structural Adjustment in Africa: Growth, Social Spending, and Debt Relief*. Center for Economic and Policy Research, Washington, DC.

Newman, D.J. and Cragg, G.M. (2012) 'Natural products as sources of new drugs over the 30 years from 1981 to 2010'. *Journal of Natural Products*, 75(3), 311–335.

Parry, B. (2004). *Trading the Genome: Investigating the Commodification of Bio-information*. Columbia University Press, New York.

Reid, W.V., Laird, S.A., Meyer, C.A., Gamez, R., Sittenfeld, A., Janzen, D.H., Gollin, M.A. and Juma, C. (1993) *Biodiversity Prospecting: Using Genetic Resources for Sustainable Development*. World Resources Institute, Washington DC.

Robinson, D.F. (2010) *Confronting Biopiracy: Challenges, Cases and International Dimensions*. Earthscan/Routledge, London.

Robinson, D.F., Drozdzewski, D. and Kiddell, L. (2014) '"You can't change our ancestors without our permission": Cultural perspectives on biopiracy.' In Fredrikksson, M. and Arvanitakis, J. (eds) *Piracy – Leakages of Modernity*. Litwin Books, Sacramento CA, pp. 55–75.

Rodney, W. (1972) *How Europe Underdeveloped Africa*. Bogle L'ouverture, London.

Said, E.W. (1978) *Orientalism: Western Concepts of the Orient*. Penguin Books, London.

Shiva, V. (1991) *The Violence of the Green Revolution*. Zed Books, New York, and Third World Network, Penang.

Smith, L.T. (1999) *Decolonizing Methodologies: Research and Indigenous Peoples*. Zed books, London.

Strathern, M. (1999) *Property, Substance, and Effect: Anthropological Essays on Persons and Things*. Athlone Press, London.

ten Brink, P. (2011) *The Economics of Ecosystems and Biodiversity in National and International Policy Making*. Earthscan/Routledge, London.

ten Kate, K. and Laird, S.A. (eds). (1999) *The Commercial Use of Biodiversity: Access to Genetic Resources and Benefit-Sharing*. Earthscan, London.

UEBT. (2010) '*A Review of Patent Activity in the Cosmetics Sector in the Context of the Ethical Sourcing of Biodiversity*'. Available at: http://ethicalbiotrade.org/dl/public-and-outreach/UEBT%20Trends%20Patents%20Activity%20Note%201%20of%204.pdf, accessed 24/4/2014.

UEBT. (2013) '*Biodiversity Barometer, 2013*'. Available at: http://ethicalbiotrade.org/dl/barometer/UEBT%20BIODIVERSITY%20BAROMETER%202013.pdf, accessed 24/4/2014.

Von Nussbaum, F., Brands, M. Hinzen, B., Weigand, S. and Häbich, D. (2006) 'Antibacterial natural products in medicinal chemistry – Exodus or revival?' *Angewandte Chemie International Edition*, 45(31), 5072–5129.

Watts, M. (2003) 'Political Ecology'. In Johnston, R.J., Gregory, D., Pratt, G. and Watts, M. (eds) *The Dictionary of Human Geography*, 4th edn. Blackwell, Oxford, p. 592.

Watts, M. and Peet, R. (2004). 'Liberating political ecology'. In Peet, R. and Watts, M. (eds) *Liberation Ecologies: Environment, Development, Social Movements*. Routledge, London, pp. 3–43.

Weber, W. and Fussenegger, M. (2009) 'The impact of synthetic biology on drug development'. *Drug Discovery Today*, 14(19/20), 956–963.

Whatmore, S. (2002) *Hybrid Geographies: Natures, Cultures, Spaces*. SAGE Publications, London.

2 The Nagoya Protocol

This chapter examines the Nagoya Protocol using two approaches in the following two sections. The first section provides a basic examination of the legal text of the Protocol, briefly discussing the purpose and likely operation of several of the articles once implemented. The second section provides a literature review, examining what a number of scholars are saying about the Protocol since it was negotiated.

The Nagoya Protocol text

The objective of the Nagoya Protocol is almost the same as the third objective of the Convention on Biological Diversity (CBD) for the 'fair and equitable sharing of benefits arising from the utilisation of genetic resources' (the full version of what gets commonly abbreviated to ABS by those working in this sphere), with the addition that ABS should thereby contribute to 'the conservation of biological resources and the sustainable use of its components' (Kamau et al., 2010).

Scope

The terminology used by the Nagoya Protocol in Article 2 provides for an expansion of scope that is significant for those in various industry sectors conducting research on natural products. The definition of 'utilisation of genetic resources' includes research and development on the genetic and/or biochemical composition of genetic resources, including the application of biotechnology. The deliberate inclusion of 'biochemical composition' was sought by many countries, particularly those in the Global South. This was pushed by those seeking to avoid narrower interpretations of the 1992 CBD text of 'genetic resources' as valuable genetic materials containing functional units of heredity (genes and their constituent DNA and RNA). The term 'biotechnology' was broadened to mean 'any technological application that uses biological systems, living organisms, or derivatives, thereof, to make or modify products or processes for specific use' – a noticeably broad and inclusive definition. This definitional expansion continues with the elaboration that 'derivative' means a 'naturally occurring biochemical compound resulting from the genetic expression or metabolism of biological or genetic resources, even if it does not contain functional units of heredity'. This expansion to include biochemical compound derivatives has the impact of rendering many products such as biological product derivatives used in skincare and cosmetics as potentially relevant under the Protocol (if R&D has been conducted). The scope of the Protocol has been broadened to include traditional knowledge (TK) associated with genetic resources in Article 3 of the Nagoya Protocol.

Access

Articles 6 and 7 deal with access to genetic resources and associated TK, respectively. While Article 6 is highly specific about certain aspects of access to genetic resources, Article 7 is much shorter and more ambiguous. Article 6 reiterates state sovereign rights over natural resources and requires that access to genetic resources for utilisation is subject to the prior informed consent (PIC) of the Party providing those resources (either the country of origin of those resources or a Party that has acquired them in accordance with the CBD). The article also adds a requirement for domestic measures which add another layer of PIC of indigenous peoples and local communities (ILCs) for access to genetic resources 'where they have the established right to grant access to such resources'. There may be relatively few countries where ILCs have clearly established legal rights over genetic resources under existing formal/codified law, but there are probably many which grant these rights under customary law (for which state recognition will vary). It is likely that individual countries and the ILCs that live there will have to examine, debate and resolve this idea of 'established rights' and this will vary from country to country. For example, many Pacific Island countries have formal legal recognitions of customary land tenure and also marine tenure. Does this also mean that those communities with land/marine tenure hold 'established rights' over the genetic resources that are found there? Many of these communities are likely to argue that they do, and this may lead to inter-community competition over genetic resources, as well as community–state tensions over 'established rights'.

To achieve Article 6 domestically, Parties are required to take the 'necessary legislative, administrative and policy measures' (suggesting some flexibility about the regulatory mechanism for doing so) to facilitate PIC (in practical terms this often means the request and provision of a permit). The written decision on PIC must be made by a competent national authority, subject to a number of stipulations about transparency, clarity, timeliness, fairness and cost-effectiveness. The issuance of a permit (or similar) indicating PIC and mutually agreed terms (MAT) then has to be notified to the ABS Clearing House (discussed later in this chapter). Article 6 also spells out the need for clear rules and procedures for establishing MAT, including among other things: a dispute settlement clause; terms on benefit-sharing including in relation to intellectual property rights (IPRs); terms on subsequent third party use, if any; and terms on changes of intent.

It is worth noting here that there are trends away from in situ bioprospecting, particularly for pharmaceuticals as discussed in Chapter 1. Indeed if we look to the recently developed EU Regulations on Access and Benefit-Sharing, the regulations leave the question of access open to member states, and they contain text on 'trusted collections' (European Commission, 2014). To be considered a trusted collection, the institution must demonstrate that it has standardised and transparent ABS procedures in place. Users who then access from a 'trusted collection' are thus assumed to have exercised their 'due diligence'. This is likely to further reinforce the sourcing of genetic resources from ex situ genebanks, which may be seen by some as positive because they are likely to have experience with ABS and standardised procedures. However, this may be seen as a negative because it seems less likely that benefit flows will get directed through to the original provider communities or countries.

The regulations also indicate that they would not be retroactive on genetic resources in those countries or as already collected and held in EU genebanks or herbariums. Early commentaries from developing countries and ILCs are likely to be critical of this, given the amount of material that has been transferred in the past. Indeed, Parry (2004) explained in detail that using technologies such as tissue cultures or chemical synthesis, and through 'microsourcing', companies are able to locate biochemical compounds of extracts from existing university departments or natural product compound libraries in many locations (including the United States as a non-Party to the CBD). She also noted that the improvement of DNA extraction techniques has allowed the use of dried or cryogenic herbarium specimens, meaning the potential sourcing of replicable genetic or biochemical materials from museums, herbariums and botanic garden collections (pp. 174–175). Indeed this has been possible since the 1980s (Rogers and Bendich, 1985). It is worth noting these trends in terms of its relevance for access and permit procedures, as countries will have to consider if/how they regulate genebanks or herbariums and collections in their territories. Certainly, botanic gardens and herbariums have been aware of these implications since the CBD (or earlier) and often require detailed ABS agreements for access to certain materials, or simple memoranda of understanding for work involving less sensitive materials (e.g. dried herbarium specimens), as Davis (2007) discussed of Kew Gardens and their Millennium Seed Bank Project. I will return to this topic in Chapter 11.

Benefit-sharing

Benefits arising from the utilisation (R&D) of genetic resources 'as well as subsequent applications and commercialisation' are to be shared with the provider Party of those resources (the country of origin or a Party that has acquired the resources in accordance with the CBD) in a fair and equitable way (Article 5). The mention of 'subsequent applications and commercialisation' suggests that once the Protocol enters into force, new uses of previously accessed and utilised genetic resources may require new benefit-sharing – the exact sequence of these aspects is something still likely to be debated (Kamau et al., 2010). In addition, those countries that are Party to the Nagoya Protocol are expected to develop measures to ensure that benefit-sharing with ILCs occurs, again where they have 'established rights' over genetic resources, and where associated TK is utilised. Benefit-sharing should be based on MAT, and there are various factors to consider here, including the power relations, timing and expertise of those involved. Indeed, some laws specify that sufficient time should be allowed such that local providers can seek appropriate legal advice (discussed in Chapter 7).

Different commentators have different ideas and expectations about benefit-sharing. The Protocol indicated that it may include both monetary and non-monetary benefits (and some of these are listed in the Annex of the Protocol), and from viewing this list it can also be seen that there are both short-term and long-term benefits that might occur. It is often important for projects to have a mixture of benefits and to mitigate any unlikely expectations of 'green gold' which rarely occur in practice. Authors such as Davis (2007) have noted the importance of short-term and non-monetary benefits such as information transfer, training and technology transfer, particularly for non-commercial research (discussed in the International Cooperative Biodiversity Groups (ICBG) cases and the Vanuatu case).

Traditional knowledge

Further to the inclusions on TK associated with genetic resources in the scope, access and benefit-sharing sections, the Protocol also deals with ILC customary laws. Parties are to consider ILC 'customary laws, community protocols and procedures' with respect to TK associated with genetic resources (this includes mechanisms to inform users) (Article 12). Parties are even encouraged to support the development by ILCs, including women, of community protocols in relation to access to TK, minimum requirements for MAT, and model contractual clauses for benefit-sharing. Although this Article is limited by the use of ambiguous language (e.g. 'as appropriate', 'in accordance with domestic law'), the requirement to consider ILC customary laws, protocols and procedures is quite a significant step for indigenous rights in international environmental law. The Article also requires Parties not to restrict the customary use and exchange of genetic resources and TK within ILCs.

As with the trend away from in situ sourcing of genetic resources, there is also a parallel trend away from sourcing associated TK in situ, particularly for commercial purposes (as evident from a number of cases discussed in the book). As with genetic resources, there is much TK that is now highly accessible because it has been disclosed or documented in the past. While this may have been permitted, there are probably many circumstances in which knowledge was collected and then entered into the public domain without the PIC of the people who held it. In many cases, TK may have always been widely held and shared, and distributed across borders and between communities. In such cases its histories and the origins of particular uses of certain genetic resources may be hard to trace (e.g. Osseo-Asare, 2014, discusses the historical use of Periwinkle in Madagascar, but also in other locations like the Philippines). In certain other cases it may be tightly held, and/or sacred knowledge. For various reasons, such as concern about disclosure prior to the start of the Santo 2006 Expedition in Vanuatu (discussed later), projects have been avoiding the collection and use of TK for R&D (I will return to this topic in Chapter 8).

However, there are still likely to be some relevant activities, such as the collection activities of botanic gardens and herbariums, and the projects of ethnobotanists. Even some current bioprospecting projects with potential commercial applications include potential sourcing from TK holders (e.g. the ICBG PNG). In any case, the mandate for clarification of the rights of ILC in relation to TK is a welcome addition to international law. The encouragement of national recognition of customary laws and protocol is also an important development in its own right.

Compliance

Articles 15 and 16 provide for compliance measures relating to ABS for genetic resources and TK, respectively. The text is quite flexible on how countries take measures to ensure that genetic resources utilised within its jurisdiction (by 'users') have been accessed in accordance with PIC and MAT of the country where they sought access (and in accordance with PIC and MAT of ILCs when TK is accessed). The Intergovernmental Committee on the Nagoya Protocol (ICNP) has been considering this issue at its meetings and a draft, heavily bracketed, text has gone forward to the third ICNP meeting in February 2014 (document: UNEP/CBD/ICNP/3/8). The main suggestion from these considerations is to have a Compliance Committee made up of members nominated by Parties and including representatives of ILCs (potentially just as observers). The Committee will promote compliance and seek to address cases of non-compliance. To date, many of the procedural aspects of the Committee's work are still under discussion. Ideas such as the establishment of an international ABS ombudsman were negotiated in the Working Group on ABS in the lead up to the final agreement of the Nagoya Protocol and are now again being discussed in the ICNP meetings. The EU Regulations on ABS have a focus on 'due diligence' by users and place a number of conditions on their use of genetic resources, including record-keeping about the samples, PIC and MAT.

Compliance with MAT is discussed in Article 18, including dispute resolution aspects that includes the encouragement of users and providers to clarify the jurisdiction in which they will follow dispute resolution processes, the applicable law(s) and options for mediation/arbitration. Parties are also required to allow for recourse under their legal systems in cases of disputes arising from MAT, which will typically be resolved through contracts law.

Monitoring

To support compliance, Parties are required to take measures to monitor the utilisation of genetic resources through the designation of checkpoints, and the encouragement of users and providers to report on the implementation of their MAT. The designation of checkpoints such as patent offices was debated in the Working Groups leading up to the promulgation of the Protocol; however, the naming of specific checkpoints was ultimately left out. Article 17 leaves considerable discretion to Parties as to what might be an appropriate check-point for their circumstances. This was a disappointment to many developing countries and groups such as the International Indigenous Forum on Biodiversity in the negotiations leading to the Nagoya Protocol, because the idea of a 'disclosure of origin' patent requirement has been sought in the WTO TRIPS Council as a remedy to patent-based biopiracy (see Orsini, 2014; Walbott et al., 2014; and Walbott, 2014 for an overview of different negotiating positions on a number of aspects of the Protocol). As a marker of innovation, due to the inventiveness requirements of patent law, this has been suggested as a useful juncture from which to identify where R&D has been conducted on genetic resources, and if PIC and MAT have been obtained (see Oldham et al., 2013).

Under the Nagoya Protocol, Parties are required to make ABS permits available to the ABS Clearing House, constituting an internationally recognised certificate of compliance (Articles 14 and 17). The certificate then serves as evidence that the genetic resource which it refers to has been accessed in accordance with PIC and MAT in the relevant country Party. The Protocol specifies that the certificate of compliance must contain the following minimum information (when it is not confidential):

a) issuing authority;
b) date of issuance;
c) the provider;
d) unique identifier of the certificate;
e) the person or entity to whom PIC was granted;
f) subject-matter or genetic resources covered by the certificate;
g) confirmation that MAT was established;
h) confirmation that PIC was obtained; and
i) commercial and/or non-commercial use.

Although the Article does not mention TK, it can be assumed that relevant PIC, MAT and subject matter detail would also be provided by the competent national authority of the provider Party.

Other aspects of the Protocol

There are several other articles in the Protocol, which warrant a brief mention. The Protocol includes a relatively weak Article on transboundary cooperation (Article 11), which encourages Parties to 'endeavour to cooperate' in situations where the same genetic resources are found across borders and/or where TK is found in more than one country. It also makes provision to establish a Global Multilateral Benefit-Sharing Mechanism to address situations where genetic resources and associated TK are transboundary, or for situations 'where it is not possible to grant or obtain PIC'. This Article was largely seen as a weak compromise after provider country positions on 'temporal scope' were largely removed from the text during negotiations. These countries were seeking some degree of retroactivity from the Protocol, whereby benefits from genetic resources accessed prior to the Protocol (and there were discussions about accessions before or after the CBD) without an ABS agreement would mean that users would be required to reach MAT retrospectively, or that new uses of previously accessed genetic resources would require users to do the same. Given that many genetic resources have been transferred in the past, this is essentially a situation where it is not possible to obtain PIC. The exact functioning of the Global Multilateral Benefit-Sharing Mechanism is still far from clear, and has been discussed on the Clearing House website portal and in Expert Working Group Meetings on ABS.

Article 8 also provides for special considerations simplified measures for ABS where research is for academic or conservation (non-commercial) purposes. It also asks Parties to consider the need for expeditious access to genetic resources in emergencies. This has been included because of the need to conduct research on viruses and bacteria during epidemics and pandemics (e.g. the avian influenza or H5N1 outbreak in 2004). Parties are also asked to consider the importance of genetic resources for food and agriculture (and food security). This has been included because of the historical and modern interdependence of countries on plant germplasm for breeding (the International Treaty on Plant Genetic Resources for Food and Agriculture – ITPGRFA – is also acknowledged in the Preamble of the Protocol). Related to this, Article 4 discusses the relationship of the Protocol with other international agreements, indicating mutual supportiveness between the Protocol and other international laws (e.g. ITPGRFA, World Health Organisation Regulations, the UN Convention of the Law of the Sea, the UN Declaration of the Rights of Indigenous Peoples, the work of the World Intellectual Property Organization and its Intergovernmental Committee on Intellectual Property, Genetic Resources, TK and Folklore – the WIPO IGC, and the WTO TRIPS Agreement).

The protocol will enter into force on 12 October 2014, after having received its 50th instrument of ratification by parties to the CBD (Article 33). As a result the first Meeting of the Parties to the Nagoya Protocol will be held in October 2014.

Literature examining the Protocol

Rather than being an exhaustive review and summary of the literature, this section seeks to highlight some of the key implementation challenges, concerns and lessons identified by relevant academic authors, policy-makers and practitioners. Since the Protocol was negotiated, many papers have been written that interpret the text from different perspectives (e.g. disciplines like law, environmental studies and economics; or from different stakeholder perspectives such as ILCs, biodiverse countries and user countries), whereas others have sought to examine its implications for countries, for specific industries or for scientific research. Given that the Protocol is just now entering into force, much of the comment remains speculative by necessity.

Stakeholder perspectives on the text

A number of papers discuss the implications of the Nagoya Protocol for research on biodiversity (mainly referring to conservation-oriented non-commercial research). During the lead-up to the Nagoya Protocol a number of researchers urged negotiators to ensure that an 'International Regime on ABS' does not inhibit non-commercial research, particularly where it is intended to benefit biodiversity conservation (e.g. Jinnah and Jungcurt, 2009). Since its development, a number of authors have commented on the need to promote early-stage research through simple procedures or contracts (Kursar, 2011), or by taking a two-stage approach to negotiation and contractual agreement. Referring to the approach of the US National Cancer Institute (NCI) and ICBGs, these authors are encouraging simplified measures for access for non-commercial research but requiring in any agreement that a change of intent towards commercial activity requires the sharing of additional benefits and/or renegotiation (see Cragg et al., 2012). This is suggested as a potential approach, particularly for provider countries implementing the Nagoya Protocol, because of the need to continue basic research on biodiversity for a range of reasons and also because of the low probability of identifying commercially viable drug candidates from bioprospecting. Kamau et al. (2010) also noted that it is the provider state's discretion as to whether to require PIC, or to allow access without prior consent. For example, despite having a national ABS framework, a de facto absence of regulations in certain states in Australia suggests the latter may be opted. This might be one way of streamlining access. Most Parties are likely to lay out more detailed access procedures including (adapted from Kamau et al., 2010) the following:

- An authorisation or permit by the competent national authority granting access and setting basic terms for access, utilisation, and benefit-sharing;
- A contract between the same body and the research institution setting basic terms for access, utilisation, and benefit-sharing;
- In relation to traditional knowledge, the PIC of its holder; and
- In relation to genetic resources, the consent of private or communal landowners or owners of the genetic resources.

Within the above process, certain aspects could be streamlined for non-commercial research. Indeed, part of the compromise here in the negotiations was that compliance measures would be introduced in user countries (Oliva, 2011). In theory, the monitoring and compliance measures would mean the ability to check if non-commercial research has changed its intent to commercial (thus potentially breaching the original permission) and to address the non-compliance.

In an article which surveys researchers who use genetic resources in the United States, the authors highlight the multiple-use practices and regular exchanges among researchers, where the line between commercial exchange and R&D is often unclear, as well as the line between commercial and non-commercial research. They focus on non-plant genetic resources and note that the numerous forms of genetic resources (e.g. microorganisms, fish eggs, oocytes, blood and semen) and means of transportation (e.g. frozen, dried and live) likely invoke a range of different rules and processes that govern their pathway of movement (Welch et al., 2013). If we consider herbarium specimens collected for taxonomic identification and comparison by botanic gardens, they are typically exchanged dried (dead) and thus arguably provide limited opportunity or likelihood of utilizing the genetic resource for commercial R&D. The authors give the example of honeybee genetic resources which may be used for different types of agricultural research – the same honeybee genetic resource might be used to study pollination efficiency, or for honey production, or for horticultural crop production (Welch et al., 2013). They note that many researchers work across sectors and national borders. Movement towards a more formal system of monitoring will have different effects on actors, pathways and outcomes – voluntary policies might be flexible but have little uptake, while more coercive rules may affect collaboration, the ability to conduct pre-competitive research or willingness to voluntarily provide non-monetary benefits (Welch et al., 2013). Thus the authors seem to be suggesting that the Nagoya Protocol will add complexity to an already complex process of exchange that is often informal, but also already involves a complex regulatory environment.

Tvedt (2013) examined patent rights and contracts through the lens of forest tree genetic resources and aquaculture contracts to provide a 'functional' understanding of genetic resources. He explained that patent claims for individually filed patents define the object of the right and how it can be utilised under the exclusivity of the holder – in essence it forms a neat and functional legal box. This can be contrasted to the CBD understanding of genetic resources, which as we can see from the previously mentioned article (Welch et al., 2013) are utilised in many ways and to which rights are rarely defined. Patent rights represent private rights for researchers, whereas much of the intent of the ABS provisions in the CBD and Nagoya Protocol are about ensuring fairness and equity for the access to essentially 'public goods' as recognised by sovereign rights over genetic resources. Thus inevitably contract rights (private law contracts or MATs) must play an important role in ensuring functionality in ABS transactions under the Nagoya Protocol, and that rights must be clearly 'defined and enumerated just as the patent system does for patents, including specifying what is not permitted after the transfer' (Tvedt, 2013, p. 140). Thus many of the transfers discussed by Welch et al. (2013) have to be clear about whether they are for commercial or non-commercial R&D (if they are for R&D at all – they might be for commodity exchange). Tvedt (2013) gave two examples: forest trees where contracts are typically used to regulate stands and regeneration of forest trees (e.g. regarding logging); in aquaculture, the sale of smolt (juvenile fish) to salmon farmers is generally regulated by contract – presumably these contracts do or should specify the intent of the use. For the Nagoya Protocol to function well, contracts regarding the exchange of genetic resources should specify if the exchange is for R&D.

A number of authors highlight that regarding ILCs it will rarely be clear who has 'established rights over genetic resources' under formal state laws (Article 6). Indigenous advocate, Harry (2011), explained that such vague language gives states broad latitude in their treatment of indigenous peoples' rights to control access, and that is likely to result in the abuse of such rights, particularly where they are not clearly defined in domestic law (also discussed in Tobin, 2013). Chennels (2013), citing Hayden (2007), warns us against the assumption that clearly defined groups exist for the sake of seeking PIC and negotiating MAT – 'legitimate political representation' and clearly defined 'state rights' are often contested within inhibitory or imposed Western legal frameworks that are the legacy of colonisation. Indeed many communities are not easily defined as 'ILCs' nor as coherent, nor might they be agreeable to negotiate an ABS agreement. Depending upon the national context, there is the question of whether a land (or sea) right would extend to an equivalent right over access to genetic resources found on that piece of land. Aside from the question of what is a legally 'established right', we can ask if one community is not willing to provide access to genetic resources, but another nearby is, who has the right to provide (or deny) access and how might benefits be shared?

Following from this, another area that has received much attention is regarding TK. Some of the above questions about genetic resources also apply to TK. Chennels (2013) discusses the challenges regarding the determination of 'TK holders'. He notes that the South African ABS legislation envisages that the 'community', defined as an ABS stakeholder, is assumed to have the capacity to assert its TK rights. Furthermore, he notes that the law specifies that the bioprospector must identify the TK holder and negotiate a benefit-sharing agreement (Chennels, 2013; see also, Myburgh, 2011). He suggests that both of these notions are optimistic. In many cases an identifiable community of TK holders might be difficult to define, nor may they be a coherently organised group, nor may they have the capacity or desire to enter into a legal agreement for benefit-sharing. Chennels (2013) also asks questions about how long the knowledge must have been held to be considered 'traditional', and whether it is still 'traditional' knowledge once it is documented, translated and transferred into the control of others. He also raises important questions about the extent to which knowledge can be exclusive and/or shared and transferred (with reference to the San-Hoodia case). This then leads us to questions about which individuals in the community (or which community) has the right to provide access and negotiate an ABS agreement? On this, Chennels (2013) discusses the transfer of knowledge regarding Pelargonium (where there is a benefit-sharing agreement between Schwabe and Xhosa communities) in South Africa that the

Xhosa may have received their knowledge from the San people. If we think further about the potential 'fluidity' of the notion of a 'TK-holder community' for the sake of an ABS agreement, it is certainly conceivable that other members of the community or communities might either want to be part of the ABS agreement or might want to reject the notion of the ABS agreement and the prospect of exclusive rights (e.g. patents) over the use of genetic resources that have TK associated with their use. This might certainly be the case in circumstances where there is a strong cultural or spiritual association with the genetic resources and associated knowledge (e.g. see Robinson et al., 2014), and where there are customary laws or protocols surrounding their use.

The inclusion in Article 12 of the requirement that Parties 'shall in accordance with domestic law take into consideration indigenous and local communities' customary laws, community protocols and procedures, as applicable, with respect to traditional knowledge associated with genetic resources' has been seen as a considerable win for the position of indigenous people (Oliva, 2011; Bavikatte and Robinson, 2011). However, the weak language surrounding the 'customary law' Article has been criticised by others (e.g. Harry, 2011). Both Tobin (2013) and Vermeylin (2013) describe in different ways the issues of wedding formal law and customary law. Vermeylin (2013) describes the place of oral history as part of customary law when used in the mainstream, and particularly, in courts where they are transcribed and translated and in doing so can be stripped of their symbolism and meaning. She notes the possibility of parts of customary law being recognised, archived and becoming part of formal law, while still in other ways creating exclusions. For example, she describes the exclusion of other San groups from the San-Hoodia ABS agreement, whose narratives against commodification found them left out of the agreement by the fetishism of exclusive property rights to exploit that genetic resource (Vermeylin, 2013).

Tobin (2013) provides a detailed examination of the challenges of introducing customary law, including customary law from a foreign jurisdiction (as might be required by compliance measures of the Nagoya Protocol and national laws) in judicial and alternative dispute resolution forums. He discusses mechanisms such as native title under common law in a number of countries, which might be leveraged so as to recognise and respect indigenous rights over TK associated with genetic resources. He, and others (Bavikatte and Jonas, 2009), also discuss the potential of biocultural protocols as a mechanism for communities to express their rights claims, customary laws, desires and intent with regards to ABS (and a range of other things, e.g., sacred sites). They might be a 'portal between customary and state law' and 'will need to be understood by national and international authorities and law' (Tobin, 2013, p. 160). For this, both Dutfield (2013) and Tobin (2013) argue the need for a shift towards an intercultural legal pluralism, whereby we stop imposing our legal solutions and start accepting the customary practices, norms and laws of indigenous communities and by ceding political space so that indigenous peoples can establish their own rules of engagement.

Special considerations and relationship with other international agreements and instruments

Several authors have examined the relationship between the Nagoya Protocol and other international conventions, different types of genetic resources (those used for food, medicines or terrestrial versus marine genetic resources) and how implementation might occur so as to avoid conflicts.

Chiarolla et al. (2013) examines the relationship between the Nagoya Protocol and the International Treaty on Plant Genetic Resources for Food and Agriculture (ITPGRFA). They note that although the bracketed text during the negotiations on ABS in Nagoya included explicit text describing the relationship between the two international laws, it was removed in favour of an article with broader application. Article 4 on the 'Relationship with International Agreements and Instruments' addresses the issue of genetic resources covered by other ABS instruments:

> Where a specialised international access and benefit-sharing instrument applies that is consistent with, and does not run counter to the objectives of the Convention and this Protocol, this Protocol does not apply for the Party or Parties to the specialised instrument in respect of the specific genetic resource covered by and for the purpose of the specialised instrument.
>
> (Article 4.4)

This means that in the context of the ITPGRFA, Party countries to it may continue using the ABS mechanism of the Treaty for genetic resources within its scope. This includes plant genetic resources held in international genebanks regulated by the Multilateral System (MLS) of the ITPGRFA and also plant species listed in the Annex of the ITPGRFA. Further to this, the preamble as well as Article 8 on 'Special Considerations' in the Protocol recognise the importance of plant genetic resources for food, agriculture and food security and that countries are 'interdependent' regarding these genetic resources. Thus the ITPGRFA strongly encourages the facilitated access to genetic resources for research and breeding. The authors also note that the benefit-sharing mechanism of the International Treaty differs from the bilateral, contract-based approach of the CBD because the benefits are shared on a multilateral basis in the MLS (Chiarolla et al., 2013). This is an important difference between the two and in the Nagoya Protocol context may again raise implications regarding considerations of 'fairness and equity' in provider countries vis-à-vis who has the established right to grant access and who is the 'TK-holder'. The ITPGRFA avoids this problem by having a centralised fund through which communities can apply for funding towards activities that conserve and develop plant genetic resources for food and agriculture. This has the effect of devolving the negotiation of benefit-sharing away from indigenous and local communities, while still allowing them to apply for grants for local breeding activities (local 'users' when we take a broader than usual view of R&D and innovation). Having a centralised fund reflects the interdependence of countries for plant genetic resources for food and agriculture. However, some might argue it will be potentially less empowering for providers than the Nagoya Protocol, which should enable ILCs to engage in ABS negotiations on their own terms. Thus, a number of 'pros and cons' for each approach can be drawn, depending upon your perspective. For example, as Walbott (2014) notes, there was some general scepticism regarding the ITPGR's ability to prevent transnational corporations from patenting genetic material accessed through the Treaty's MLS. On the other hand, it has a Standard Material Transfer Agreement which removes the need for MAT to be negotiated in every case, simplifying the process.

If we consider the implications of the Nagoya Protocol for health, there are some important potential impacts on access to medicines stemming from access to genetic resources – especially for medicines such as vaccines. Wilke (2013) noted that the ABS implications of public health emergencies became most apparent when the H5N1 or avian flu crisis began, whereby Indonesia halted supply of virus samples to the WHO because of concerns about attempts by an Australian company to obtain a vaccine patent. Because of this sort of scenario, Article 8.b of the Protocol calls upon States to consider special rules on access and benefit-sharing, including expedited access to affordable treatments for those in need, during health emergencies. As Wilke (2013) explained, the scope of Article 8.b is broad, covering all different kinds of health emergencies for humans, animals and plants and all related genetic resources – whether they are of pathogenic or non-pathogenic nature. However, she does note that the text is rather ambiguous and the language is weak, leaving much discretion to countries as to how they implement special considerations of ABS for public health emergencies. As a consequence, these ambiguities may impact the speed and effectiveness of access to genetic resources, if countries do not appropriately address this provision.

Many questions have also been raised about the scope of the Nagoya Protocol to areas beyond national jurisdiction. This was discussed until the dying hours of the negotiations on the international regime in Nagoya, and until close to the end there was bracketed text on access to genetic resources outside the limits of national jurisdiction. Ultimately, Article 3 (which refers to Article 15 of the CBD emphasising sovereign rights, and limits its language to access from Contracting Parties) and the text in Article 6 on access clearly refers to sovereign rights and consent of the Party providing genetic resources. Article 10 on the Global Multilateral Benefit-Sharing Mechanism instead requests Parties to consider the need for such a mechanism to deal with access to genetic resources and associated TK 'that occur in transboundary situations or for which it is not possible to grant or obtain PIC.' This might include access to genetic resources in transboundary marine areas, on the high seas, or in locations like Antarctica (see Leary, 2009). As noted earlier in this chapter, the full modalities of this mechanism are being worked out in the current ICNP meetings. In the meantime a number of authors have raised some interesting questions about marine genetic resources and different areas of jurisdiction.

Although the 1982 United Nations Convention on the Law of the Sea (UNCLOS) does not explicitly mention genetic resources or 'access', it does contain numerous mentions of marine life and organisms, and it also regulates marine scientific research. In a detailed paper, Salpin (2013) examined the two international agreements, noting:

> . . . the regime for ABS under the Nagoya Protocol does not take into account the complexity of the jurisdictional framework under the law of the sea, where the maritime space is multi-dimensional and divided both horizontally and vertically in a number of zones within which the rights of the coastal State and other States vary. There are thus different sets of regimes under national jurisdiction, depending on where an activity takes place: in the internal waters, archipelagic waters and the territorial sea; or in the exclusive economic zone (EEZ) and the continental shelf.
>
> (Salpin, 2013, p. 154)

Thus, in practice there may be different national agencies, community groups and other stakeholders with different responsibilities over the management and access to resources from these zones. There may be situations whereby customary rights extend into different marine zones. Hickey (2006) has examined these in Vanuatu, and the way communities regulate the catch of certain species at certain times of the year, for example, and for which there are different roles for both communities and agencies (e.g. Department of Fisheries) (see also Vierros et al., 2010, for further examples from the Pacific). Regarding the 'utilisation of genetic resources', questions have been raised regarding the potential for change of intent in cases where, for example, companies have extracted minerals from the seabed of the territorial waters or EEZ of a country, or where fish are being caught commercially in a country's waters. The interest in and potential for utilisation of genetic resources (meaning R&D on their genetic and/or biochemical composition) found among those minerals or derived from commercial fish stock has increased in recent years (see Leary, 2007). This brings us back to Tvedt's (2013) points about the functionality and role of contract law, for which relevant national authorities will now have to further consider the implications of ABS and a possible change of intent arising from commercial extractions of natural resources. Many of these responsibilities regarding ABS will vary between countries and will have to be worked out domestically, where possible, or bilaterally in other cases, or else through the eventual establishment of the Global Multilateral Benefit-Sharing Mechanism.

Regarding marine areas beyond national jurisdiction, as Salpin (2013, p. 175) noted, the 'high seas are open to all States, and freedom of the high seas, which includes the freedom of fishing and of scientific research, must be exercised under the conditions laid down by UNCLOS and by other rules of international law'. Under UNCLOS, there are also explicit requirements for equitable sharing of financial and other economic benefits derived from exploration activities and exploitation of resources of the seabed and ocean floor beyond areas of national jurisdiction (the 'Area' – which is recognised as the common heritage of humankind). All states also have the right to conduct marine scientific research on the high seas and the Area, and UNCLOS also encourages cooperation and benefit-sharing relating to this research, particularly technology transfer and non-monetary benefits (e.g. publications and data sharing) (Salpin, 2013). However, there is still some contention between different countries as to whether principles in UNCLOS (including benefit-sharing) laid down regarding the regulation of the Area of seabed beyond national jurisdictions also apply to marine genetic resources or not. This is being further discussed in a Working Group of the General Assembly of the United Nations.

References

Bavikatte, K. and Jonas, H. (eds) (2009) *Bio-Cultural Community Protocols: A Community Approach to Ensuring the Integrity of Environmental Law and Policy.* Natural Justice, Cape Town, and Nairobi: UNEP, Nairobi.

Bavikatte, K. and Robinson, D. (2011) 'Towards a people's history of the law: Biocultural jurisprudence and the Nagoya Protocol on access and benefit sharing'. *Law, Environment and Development Journal,* 7(1), 35–51.

Chennels, R. (2013) 'Traditional knowledge and benefit sharing after the Nagoya Protocol: Three cases from South Africa'. *Law, Environment and Development Journal,* 9(2), 163–184.

Chiarolla, C., Louafi, S. and Schloen, M. (2013) 'An analysis of the relationship between the Nagoya Protocol and instruments related to genetic resources for food and agriculture and farmers' rights'. In Morgera, E., Buck, M. and Tsioumani, E. (eds) *The 2010 Nagoya Protocol on Access and Benefit-sharing in Perspective Implications for International Law and Implementation Challenges.* Martinus Nijhoff Publishers, Leiden, pp. 123–128.

Cragg, G.M., Katz, F., Newman, D.J. and Rosenthal, J. (2012) 'The impact of the United Nations Convention on Biological Diversity on natural products research'. *Natural Products Reports,* 29(12), 1407–1423.

Davis, K. (2007) 'Biodiversity, botanical institutions and benefit sharing: Comments on the impact of the convention on biological diversity'. In McManis, C. (ed) *Biodiversity and the Law: Intellectual Property, Biotechnology & Traditional Knowledge.* Earthscan, London, pp. 71–76.

Dutfield, G. (2013) 'Transboundary resources, consent and customary law'. *Law, Environment and Development Journal,* 9(2), 259–263.

European Commission. (2014) '*Access and Benefit Sharing*'. Available at: http://ec.europa.eu/environment/biodiversity/international/abs/, accessed 29/4/2014.

Harry, D. (2011) 'Biocolonialsim and indigenous knowledge in United Nations discourse'. *Griffith Law Review,* 20(3), 702–727.

Hayden, C. (2007) 'Taking as giving. Bioscience, exchange, and the politics of benefit sharing'. *Social Studies of Science,* 37(5), 729.

Hickey, F. (2006) 'Traditional marine resource management in Vanuatu: Acknowledging, supporting and strengthening indigenous management systems'. *SPC Traditional Marine Resource Management and Knowledge Information Bulletin,* 20, 11–23.

Jinnah, S. and Jungcurt, S. (2009) 'Could access requirements stifle your research?' *Science,* 323(5913), 464–465.

Kamau, E.C., Fedder, B. and Winter, G. (2010) 'The Nagoya Protocol on access to genetic resources and benefit sharing: What is new and what are the implications for provider and user countries and the scientific community?' *Law, Environment and Development Journal,* 6(3), 246–262.

Kursar, T.A. (2011) 'What are the implications of the Nagoya Protocol for research on biodiversity?' *BioScience,* 61(4), 256–257.

Leary, D.K. (2007) *International Law and the Genetic Resources of the Deep Sea*, 1. Martinus Nijhoff Publishers, Leiden, The Netherlands.

Leary, D.K. (2009) 'Bioprospecting in Antartica and the Arctic. Common challenges?' *The Yearbook of Polar Law*, 1, 145–174.

Myburgh, A.F. (2011) 'Legal developments in the protection of plant-related traditional knowledge: An intellectual property lawyer's perspective of the International and South African Legal Framework'. *South African Journal of Botany*, 77, 844–849.

Oldham, P., Hall, S. and Forero, O. (2013) 'Biological diversity in the patent system'. *PLOS One*, 8(11), 1–16.

Oliva, M.J. (2011) 'Sharing the benefits of biodiversity: A new international protocol and its implications for research and development'. *Planta Medica*, 77(11), 1221–1227.

Orsini, A. (2014) 'The role of non-state actors in the Nagoya Protocol negotiations'. In Oberthür, S. and Rosendal, G.K. (eds) *Global Governance of Genetic Resources: Access and Benefit-Sharing after the Nagoya Protocol*. Routledge, Oxon.

Osseo-Asare, A.D. (2014) *Bitter Roots: The Search for Healing Plants in Africa*. University of Chicago Press, Chicago.

Parry, B. (2004). *Trading the Genome: Investigating the Commodification of Bio-Information*. Columbia University Press, New York.

Robinson, D.F. Drozdzewski, D. and Kiddell, L. (2014) ' "You can't change our ancestors without our permission": Cultural perspectives on biopiracy'. In Fredrikksson, M. and Arvanitakis, J. (eds) *Piracy – Leakages of Modernity*. Litwin Books, Sacramento CA, pp. 55–75.

Rogers, S.O. and Bendich, A.J. (1985) 'Extraction of DNA from milligram amounts of fresh, herbarium and mummified plant tissues'. *Plant Molecular Biology*, 5(2), 69–76.

Salpin, C. (2013) 'The law of the sea: A before and an after Nagoya?' In Morgera, E., Buck, M. and Tsioumani, E. (eds) *The 2010 Nagoya Protocol on Access and Benefit-Sharing in Perspective Implications for International Law and Implementation Challenges*. Martinus Nijhoff Publishers, Leiden, pp. 149–184.

Tobin, B. (2013) 'Bridging the Nagoya compliance gap: The fundamental role of customary law in protection of indigenous peoples' resource and knowledge rights'. *Law, Environment and Development Journal*, 9(2), 142–162.

Tvedt, M.W. (2013) 'Disentangling rights to genetic resources illustrated by aquaculture and forest sectors'. *Law, Environment and Development Journal*, 9(2), 127–141.

Vermeylin, S. (2013) 'The Nagoya Protocol and customary law: The paradox of narratives in the law'. *Law, Environment and Development Journal*, 9(2), 185–201.

Vierros, M., Tawake, A., Hickey, F., Tiraa, A. and Noa, R. (2010) *Traditional Marine Management Areas of the Pacific in the Context of National and International Law and Policy*. United Nations University – Traditional Knowledge Initiative, Darwin, Australia.

Walbott, L. (2014) 'Goals, strategies and success of the African group in the negotiations of the Nagoya Protocol'. In Oberthür, S. and Rosendal, G.K. (eds) *Global Governance of Genetic Resources: Access and Benefit-Sharing after the Nagoya Protocol*. Routledge, Oxon, pp. 114–141.

Walbott, L., Wolff F. and Pozarowska, J. (2014) 'The negotiations of the Nagoya Protocol: Issues, coalitions and process'. In Oberthür, S. and Rosendal, G.K. (eds) *Global Governance of Genetic Resources: Access and Benefit-Sharing after the Nagoya Protocol*. Routledge, Oxon, pp. 33–59.

Welch, E.W., Shin, E. and Long, J. (2013) 'Potential Effects of the Nagoya Protocol on the exchange of non-plant genetic resources for scientific research: Actors, paths and consequences'. *Ecological Economics*, 86, 136–147.

Wilke, M. (2013) 'A Healthy Look at the Nagoya Protocol-Implications for Global Health Governance' in Morgera, E., Buck, M., and Tsioumani, E. (eds) The 2010 Nagoya Protocol on Access and Benefit-sharing in Perspective Implications for International Law and Implementation Challenges. Martinus Nijhoff Publishers, Leiden, pp. 123-148.

Part II

Biodiscovery, access and benefit-sharing case studies

3 The International Cooperative Biodiversity Groups Madagascar

The ICBG program

The International Cooperative Biodiversity Groups (ICBG) Program was established in 1992 following a conference involving the US National Institutes of Health (NIH), the National Science Foundation and the US Agency for International Development (USAID), which focused on the potential relationships between drug development, biological diversity and economic growth. The ICBG program aims to

> ... integrate improvement of human health through drug discovery, creation of incentives for conservation of biodiversity, and promotion of scientific research and sustainable economic activity that focuses on environment, health, equity and democracy.
>
> (ICBG, accessed 21/8/13)

The program emphasises natural products drug discovery (or bioprospecting) as a way to 'promote scientific capacity development and economic incentives to conserve the biological resources from which these products are derived' under appropriate circumstances. It therefore closely adopts the underlying logic for the access and benefit-sharing (ABS) provisions stated in 'Article 1, Objectives' and text of the CBD.

The ICBG is financed through 5-year grants from 3 US government organisations: the NIH, the National Science Foundation and US Department of Agriculture (USDOA).[1] This funding was channelled into a single grant delivered by the Fogarty International Centre (FIC) of the NIH to the recipients of the ICBG grants, and then from 2008, two separate awards were made from NIH and USDOA. The ICBG grant recipients are made up of consortia of researchers and partners in the United States and in biodiverse foreign low- and middle-income countries (as defined by World Bank criteria) – the 'provider' countries of genetic resources, samples or extracts from those genetic resources. The consortia also include companies working in natural products drug discovery, agricultural biotechnology or biochemistry and related scientific fields. The ICBG program therefore has an international, public–private partnership approach. It supports 'interdisciplinary research teams in the exploration and discovery of novel compounds and extracts from nature with potential for development as therapeutic agents for multiple disease targets while building research capacity in partnering countries' (FIC, accessed 27/9/2013). Currently, active grants also include a focus on biodiscovery for agricultural, crop protection and animal health applications as well as bioenergy agents and biofuels.

The 'interdisciplinary' research teams often include conservation biologists and sometimes social scientists to contribute to the other main goals of the ICBG: 'to conserve biodiversity through valuation of natural resources, training and infrastructure building to aid in management; to promote sustainable economic activity of communities, primarily in less developed countries in which much of the world's biodiversity is found' (Rosenthal, 1997). The ICBG projects typically involve teams of researchers based in the United States as well as teams of researchers, facilitators and project staff in the host country (these may be from universities, NGOs or government departments), working together to meet the multiple objectives of the program. In this chapter we discuss the ICBG Madagascar, where interviews were conducted relating to the ABS agreements in place in 2013. Later, in Chapters 9 and 10 some other ICBG projects (Panama and Papua New Guinea, in particular) are discussed.

The ICBG Madagascar

The ICBG Madagascar is led by Dr David Kingston of the Virginia Polytechnic Institute and State University in Blacksburg, Virginia. Although Dr Kingston's first phase of ICBG funding began in Suriname, this case study focuses on the following 15 years of ICBG funding in Madagascar from 1998 to 2013 (with the project reaching the end of its funding cycle recently in August 2013). The ICBG Madagascar consortium is made up of the following partners:

- Virginia Polytechnic Institute and State University in Blacksburg, Virginia;
- Madagascan Centre Nationale d'Application et des Recherches Pharmaceutiques (CNARP);
- Madagascan Centre Nationale de Recherches sur l'Environment (CNRE);
- Madagascan Centre Nationale de Recherches Oceanographiques (CNRO);
- Missouri Botanical Garden (MBG);
- Conservation International (CI);
- Eisai Pharmaceutical Research Institute; and
- Dow Agrosciences (DAS).

Due to the unique climate, geological structure and biodiversity of Madagascar, it provides a promising site for bioprospecting unique biological samples. Extracts from Madagascan biological materials (plants, soil microbes and marine organisms – mainly micro-organisms) are sent to laboratories in both Antananarivo and the United States to be tested in assays for effects on cancer, immunology, malaria, neurological disease, tuberculosis and agrochemistry. In the last award cycle, DNA from micro-organisms was also analysed for taxonomic and other purposes.

Within ICBG projects there are objectives for biodiversity conservation and also often local economic development. Through the in-country facilitation by MBG and Conservation International, the ICBG has made several contributions in this regard (discussed below). The Madagascar ICBG secured considerable upfront funds from its industrial partners at first Bristol-Myers Squibb (no longer part of the ICBG consortium) and DAS and then Eisai Research Institute (now Eisai, Inc.) which have been utilised to provide amenities for the communities adjacent to the bioprospecting areas. In addition the ICBG has funded a range of biological inventory and conservation-related activities in these areas.

As indicated by Prof. David Kingston, 2 contracts establish the consortium and governance of the ICBG Madagascar:

> Specifically, there are two "all-party Research Agreements" that govern the roles of each party, the IP rights of the various parties, the royalty rates payable in the event of a licensed product, and so on. The two agreements were so that Dow AgroSciences (DAS) and Eisai could keep proprietary information confidential from the other company. The agreements also committed DAS and Eisai to annual "Upfront Compensation" payments; the relevant clause relating to these payments read in part "All Parties agree that these funds will be used only for research, research training, research infrastructure, biodiversity conservation, or economic development in Madagascar, in consultation with local stakeholders." All parties signed these agreements.
>
> (pers. comm. 22/10/2013)

As with most commercial contracts of this nature, these are confidential and so the full details cannot be disclosed. Most ABS agreements contain details about several things. Specific to this agreement the following are relevant:

- The type of genetic resources – plant, marine and microbial extracts (DNA and biochemical compounds);
- The intended utilisation of the genetic resources – scientific as well as potential commercial purposes;
- The relevant sectoral utilisation – for crop protection and herbicidal use (DAS) and pharmaceutical use (Eisai, and earlier Bristol-Myers Squibb);
- The use of traditional knowledge – not utilised in this case;
- Whether third party transfers are allowed – restrictions were placed here for commercial reasons and so as to gain maximum benefit for the members of the consortium. What is unclear is if third party transfers might occur at a later date in the future, and what terms the contract has relating to this;
- Intellectual property rights – from interviews it seems clear that the researchers who make a genuine discovery have asserted their rights to patent this invention;
- Upfront payments – sizable payments were agreed to for each 5 year phase;

- Milestone payments – at key developmental intervals;
- Royalty payments – once a drug or product receives marketing approval and reaches the market.

With its first phase of funding in Madagascar in 1998–2003, the project was based in areas surrounding the Zahamena National Park in the central-east of the country and involved CNARP, as well as MBG which vouchered herbarium specimens during this phase. This project phase focused on collection of plants. The second round of ICBG support in 2003–2008 saw the project move to the dry Diana region of Northern Madagascar. Here, MBG was involved in vouchering and identifying species in an array of dry forests such as those at Oronjia and Montagne des Français, with CNARP taking samples for preparation for analysis and replicates collected by MBG. Also in this region at Ambodivahibe, Conservation International received funding from the ICBG for the facilitation of community projects and for assistance with the protection of locally managed marine conservation areas. Bioprospecting of marine micro-organisms was undertaken in or near these locally managed marine conservation areas. The third phase saw the project move to soil collection activities for the isolation of microbial species in the central region of Madagascar at 5 sites where MBG was already supporting conservation projects under the ICBG, as well as continued bioprospecting and activities in the northern Diana Region, and marine collections at various sites around the island. These are explained in more detail below.

Access and R&D conducted

Madagascar has not had, and still does not have a formal policy on bioprospecting or ABS. There is currently a draft ABS law that has been developed by a quasi-governmental agency Service d'Appui a la Gestion de l'Environnemont (SAGE) (Naritiana, interview, 2/9/2013). Madagascar has signed the Nagoya Protocol in September 2011 (it ratified the CBD in 2006) and is preparing to ratify it with the assistance of the GIZ-led multi-donor ABS Capacity Development Initiative. Given the current political situation in Madagascar, following what many countries perceived as a coup in 2009, it has been difficult to make progress on matters such as this. The National Focal Point on ABS indicated that they hoped to progress the ratification process once the country had gone to Presidential elections (which were held in late 2013), with a draft law almost ready to be considered in parliament.

Despite the absence of a formal national law or policy, there have been numerous bioprospecting activities in Madagascar – some with and some without explicit authorisation it seems (Osseo-Asare (2014)). In November 1999, a research proposal covering ICBG-Madagascar's activities submitted jointly by CNARP, CI and MBG was approved by the Malagasy Government represented by the National Office for the Environment (ONE) and in December 1999 the first field trip of the project was conducted at Zahamena (WLBC/MBG, accessed 17/10/2013).

One of the Madagascan partners, Dr Ratsimbason from CNARP, indicated that they had been reluctant about the ICBG when they were first approached because the United States has not ratified the CBD (and now as a CBD non-Party had not signed the Nagoya Protocol). However, they were surprised that the ICBG was intent on applying the spirit of the CBD – instead through a contractual agreement which reflected key CBD principles. Dr Ratsimbason indicated 'this is exactly the situation here in every aspect' indicating their appreciation that the ICBG had aligned with the CBD principles after 15 years of activities in Madagascar. He also described the 'beautiful work' that was done by MBG and CI with communities with the upfront compensation funds and ICBG conservation funding (described below) (interview, September 2013).

Regarding preliminary local access to bioprospecting sites, MBG staff noted that prior to all of the collection activities a public forum would be held with all of the community invited, including mayors and village heads. At these meetings the MBG facilitators discuss the intent of the project and to gain community permission to conduct the bioprospecting and vouchering for herbarium (dried) specimens (i.e. prior informed consent) and to agree on the precise location of the sampling with the community. They then discuss with the community the potential options for the use of upfront compensation funds – the community provides considerable input while MBG or CI facilitators seek to ensure the maximisation of the number of beneficiaries and that the beneficiaries are indeed the community rather than just specific individuals. This is discussed further in the section on benefit-sharing.

Importantly, the project did not include traditional knowledge within the scope of the agreement. One interviewee noted that due to some challenges in the ICBG Suriname, also led by Prof. Kingston, and because of some concerns raised by Malagasy stakeholders (including traditional healers) in the early stages of project development, it was decided that traditional knowledge would not be collected and documented. Neimark (2012) also noted that there was dissatisfaction with the speed with which ethnobotanical explorations gave leads for drug discovery research in the ICBG Suriname case and so it was perceived as a less productive approach for the Madagascan ICBG project.

The first phase

The first phase of biodiscovery activity involved collection of plant samples from unclassified and classified forests adjacent to the Zahamena Protected Area (bioprospecting is not allowed within Madagascar's protected areas). In addition, botanical inventories were conducted in the area. Zahamena is located on the eastern side of the country, approximately 175 km northeast of Antananarivo in an area that is difficult to access, requiring 1 day of road travel and up to 3 days hike to reach the collection sites. The 2 protected areas there together have an area of 64,400 ha, and several unique environmental characteristics (climate, topography and vegetation types) make this an interesting site for bioprospecting (WLBC/MBG, accessed 17/10/2013). Initially, activities at this site were facilitated by CI.

Due to the inaccessibility of the site, up to 15 porters were hired to transport camping materials, provisions and collecting equipment from the vehicle to the campsite and also to transport collections back to the vehicle. Fieldwork teams were typically able to collect samples from the forest within a 5 km radius of the camp. At the start and end of all fieldtrips, courtesy visits were made to the Head of the Village and to local CI representatives (WLBC/MBG, accessed 17/10/2013). Although this site was not visited during fieldwork for this case study (due to its inaccessibility), interviews and discussions with MBG staff who were involved in the collection were conducted.

Typically the fieldwork team would be composed of 1 botanist from MBG, 2 botanists from CNARP and 1 botanist and 1 guide from CI (WLBC/MBG, accessed 17/10/2013). However, from my interviews with MBG staff, it was explained that there were sometimes additional Malagasy botanists or botanists in training from MBG who attended and contributed to the collection activities (e.g. Jeremie Razafitsalama received training while working in this first phase in Zahamena and is now the *Chef de Projet* for MBG in Oronjia).

Samples for bioassay were collected from fertile plants in forests adjacent to the protected area. One kilogram wet weight of each plant part would be collected, except for bark for which only 0.5 kg would be collected. A typical woody tree species would yield 4–6 different samples (e.g. roots, wood, bark, twigs, leaves, flowers, fruit and seeds), while other plants might yield just 1 or 2 plant parts (WLBC/MBG, accessed 17/10/2013). Each sample collected for biodiscovery purposes is labelled with a unique code that allows the sample to be linked to a vouchered herbarium specimen. The specimen was made in 6 replicates so that vouchers could be lodged in Madagascar's national herbarium and also with several international herbaria and with the specialist for the taxa concerned. The herbarium specimens included detailed field notes that included precise location information allowing the species to be identified and located again in the field should biomedical/biochemical assays using the parts of the species prove interesting. It was noted in interviews that if ethnobotanical information was incidentally conveyed to the botanists, it is not included in their documentation of their inventories in the TROPICOS database that MBG administers for public and conservation benefit (Birkinshaw, interview, 2/9/2013). It is also worth noting that if only limited voucher samples can be collected (due to relative rarity of an individual of a species) then they would first be provided to the National Herbarium of Madagascar at Park Tsimbazaza in Antananarivo (i.e. keeping a specimen in-country is of the highest priority and is required by the Malagasy authorities) (Birkinshaw, 2/9/2013). MBG indicates that samples are not collected from species which are considered rare at the site or species which are listed in the appendices of CITES (WLBC/MBG, accessed 17/10/2013).

The second phase

The second phase of the project saw ICBGs activities shift to the dry forests of northern Madagascar. Bioprospecting was undertaken at several locations in the North. These included the following: Mt des Francais, Daraina, Andavakoera, Oranjia, Sahafary, Ambilobe, Cap d'Ambre, Mts d'Ambre, Ambolobozokely and Ambolobozobe. Due to time constraints, interviews in the field for this section of this chapter were limited to locations around Mt des Francais and Oronjia. The Oronjia peninsula lies to the east across the bay from the city of Antsiranana and is the location of most of the ICBG activities in the region. At the base of the peninsula is the Montagne des Français – a limestone massif, in the centre of the peninsula up to the Northern tip near Ramena is a dry deciduous forest, and to the east lies Ambodivahibe, all of which were areas where research was conducted. Although there was some sampling of plant parts in the region by the CNARP researchers, this shifted to soil sampling (by CNRE) and marine sampling for micro-organisms (by CNRO) at a later date. This was coupled with vouchering of replicate specimens of plant parts as well as botanical inventory research to identify priority areas for plant conservation by MBG. In the earliest stage, SAGE was involved in this site; however, this was taken over by MBG. CI has had a considerable role in implementing marine protected areas – continuing until 2013 with the support of ICBG (probably even into 2014 with residual conservation funds). The implementation of upfront compensation in villages and terrestrial conservation activities in the forests here was taken over by MBG. The upfront compensation was still being spent in communities in 2013 and may continue for several months into 2014.

Herbarium collections were also conducted in this phase. Chris Birkinshaw from MBG indicated that about 5,000 specimens had been collected, and these species identified in the first 2 phases of the ICBG project. Many new species were identified as a result of these activities, and distribution maps have been developed and included in the TROPICOS database, being useful for planning and decision-making for the conservation of these species (interview, 2/9/2013). These were collected using the techniques described above. From interviews conducted in the Diana region, people from the relevant villages commented that they had been consulted about the collection activities prior to the research and were often aware that the upfront compensation was in return for access to conduct the research. When asked if they had concerns about the research none were raised. However, they were generally unclear about the results of the research and many expressed an interest in knowing more about what had happened since the researchers had come and collected (interviews, 7/9/2013 and 9/9/2013).

The third phase

During the third phase, the activities of the ICBG continued in the Diana region at and near the Oronjia peninsula. However, additional bioprospecting activities, conservation activities, vouchering and specimen identification and community development projects were undertaken at 4 other sites in Madagascar: at Ibity Massif, Ankafobe, Vohibe forest (Ambalabe) and Pointe à Larrée. For this chapter, fieldwork was conducted at 3 of these sites: Ibity, Ankafobe and Oronjia (noted above). During the third phase of bioprospecting, only marine organisms and soil microbes were collected for extraction and analysis. Plants were not part of the bioprospecting; however, botanical inventory was conducted around the soil sampling sites as part of a research project that aimed to relate soil microbial composition and diversity with botanical composition and diversity at sites such as Ibity and Oronjia (Mamisoa, 5/9/13). These 5 bioprospecting sites were chosen by the ICBG consortium because they offer considerable contrasts in geology, soil types, microclimates and local ecological conditions (Birkinshaw, 2/9/2013). Interviews again revealed a general awareness that the upfront compensation was a result of providing research access to local sites and there were no concerns raised about this. Rather the interviewees were curious about what had occurred since the fieldwork was conducted (interviews, 5/9/2013). Figure 3.1 highlights the locations of MBG's conservation projects, most of which were also bioprospecting locations.

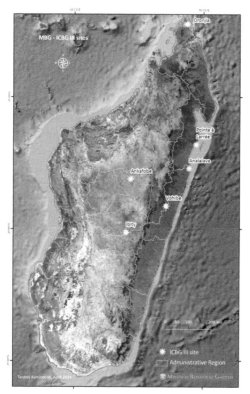

Figure 3.1 Location of MBGs conservation projects and a number of bioprospecting sites

Source: MBG Madagascar.

Research and development

To date, the ICBG Madagascar has not been successful in developing a drug to the point of clinical trials (nor has any other ICBG project). However, Kingston (2011) noted the potential of a number of compounds at varying stages of development. Specifically from Madagascar he described a derivative (schweinfurthins) isolated from the Madagascan plant *Macarangaalnifolia*, which has potent antiproliferative activity, meaning that it might be a promising candidate for development towards a treatment for cancer (Kingston, 2011).

As of 2013 there are more than 61 peer-reviewed journal papers (from a total of approximately 75 with relevance to Madagascar) and 3 books which appear to include Malagasy authors from the ICBG Madagascar project, indicating in quantitative terms the strength of the collaboration between the partners in the United States and in Madagascar.[2] While the majority of these articles focus on various developments in drug discovery, at least 17 articles and the 3 books have a biodiversity and conservation focus, mainly focused on taxonomy and identification of new species and threats to these species. A quick survey of the drug discovery articles indicates that the majority of articles focus on antiproliferative and cytotoxic compounds, suggesting a heavy research focus on agents that might inhibit tumour growth. Of greater direct relevance to Madagascar, there are also a small number of papers from the ICBG focused on antimalarial/antiplasmodial activity (Ratsimbason et al., 2009; Cao et al., 2006; Rakotonjatovo et al., 2006). In addition to the peer-reviewed papers, there are many other conference papers, proceedings publications and book chapters also involving or acknowledging Malagasy scientists on a range of related topics.

Monetary benefit-sharing

Due to the size and complexity of this project, I will discuss several monetary, potential monetary and non-monetary benefits in separate sections (summarised in Table 3.1).

The total upfront funding over the last 5 years was $300,000, but more than this was spent over the last 5 years because of unspent funds from previous years. Similar amounts were provided in each 5 year period – however, interviewees noted that not all of the money had been spent in those phases and some had been carried over into recent projects in 2013.

Upfront compensation funding is provided by the industry partners to the ICBG. The upfront funds went into a bank account in the name of the ONE. This is then split such that 50 per cent of the funding goes to the national centres for specific infrastructure enhancement and purchase of research equipment, plus 50 per cent to community-based projects around the bioprospecting collection sites. All parties involved committed to the use of these funds only for research, research training, research infrastructure, biodiversity

Table 3.1 Monetary benefits from the ICBG Madagascar

Monetary benefit	Beneficiaries
Upfront compensation funds from industry.	• 50% to Madagascan National Research Centres. • 50% to communities in Madagascar where bioprospecting was undertaken.
Funding for conservation projects from the Fogarty International Centre (FIC) of the NIH which administers the ICBG program.	• Administered locally by MBG and CI for local conservation benefits, and local community benefits.

Potential monetary benefit	Beneficiaries
Milestone payments at specific development intervals.	• 50% to Madagascan National Research Centres. • 50% to communities in Madagascar where bioprospecting was undertaken.
Royalties upon marketing approval and release of the products to market.	• 50% to Madagascan National Research Centres. • 50% to communities in Madagascar where bioprospecting was undertaken.

conservation or economic development in Madagascar, in consultation with local stakeholders. The funds were managed by a group of local stakeholders, including representatives from CNARP, CNRE, MBG, CI and CNRO who established a (Madagascan) Stakeholder Committee. Prof. Kingston, NIH and the 2 companies involved reputedly had no control over how the money was spent (Kingston, pers. comm. 22/10/2013). Decisions on the use of the funds were made by the Stakeholders Committee. Grants were awarded from the 50 per cent allocated towards community-based projects, based on an assessment by the Steering Committee of the proposals from the local populations (Kingston, pers. comm. 22/10/2013). After being met by members of the Madagascan Stakeholder Committee to explain the project, the community had the option of accepting or refusing the researchers. The community would have to agree to allow research in the areas, and then subsequently they would have to propose community development proposals to receive a portion of the upfront compensation funds (Birkinshaw, pers. comm. 2/9/13).

Government upfront compensation funds

The 'government' funds were shared between the 3 government laboratories – CNARP, CNRE and CNRO. Since the 3 institutes agreed to share the Madagascar government centres portion of the upfront funds equally, Prof. Kingston estimated that they each had spent about $140,000 over the last 5 years – this included the government share of unspent upfront funds from earlier years (Kingston, pers. comm. 22/10/2013). The following expenditures were reported:

- Expenditures of $138,899 by CNRE for the purchase of scientific equipment. Much but not all of this equipment was for ICBG-related work. Some was for equipment to support other work at CNRE.
- Expenditures of $123,800 by CNARP for the purchase of scientific equipment. This was primarily for a high performance liquid chromatograph (HPLC) for ICBG studies and a gas chromatograph mass spectrometer (GC-MS) for non-ICBG studies.
- CNRO spent about $50,000 on the purchase of a 4WD vehicle to assist in marine collection activities around the country. They also purchased diving equipment for their marine collections, and set up a new marine research station at Vangaindrano in SE Madagascar. Total expenditure is likely to be close to $140,000.

Dr Michel Ratsimbason at CNARP discussed the ICBG project in broad terms as well as the specific of the upfront compensation payments and their impacts. After indicating the many benefits for Madagascar (described earlier) he noted that no project is perfect – that CNARP originally did data collection preparation and extractions from plants then sent these to the United States: 'For CNARP staff this is not very useful' (interview, 4/9/2013). He pointed out that there had been some criticism of the ICBG because of its focus on cancer research and pharmaceutical research, wherein:

> In industrial pharmaceuticals the owner of the intellectual property is the one who identifies the compound. Unfortunately no identification of structures has occurred in laboratories in Madagascar. If found . . . the laboratory [who owns the patent] would be the one that identified the structure, so this is more likely to be in the US.
>
> (interview, 4/9/2013)

Therefore CNARP had asked for the US researchers to help to set up a malaria research laboratory at one point earlier in the ICBG phases, to do extraction and purification work. After some negotiations the ICBG consortium agreed and bought some apparatus for conducting chemical research:

> CNARP has focused on plant extracts. We have collected about 4500 plant extracts from parts of plants (leaves, bark, roots – 5 parts). Half have been tested on malaria at CNARP and half are remaining. There is a chloroquinone resistance problem here so we test of the strain that is resistant. In the US they have a thematic test and this points at which resistance occurs in the *Plasmodium* at the molecular level – at Virginia they do this. Maryland will also send samples to CNARP for malaria testing, from extracts from actinomycetes or other microbiological extracts. This is a good collaboration and the extraction is also conducted at CNRE – this is the only laboratory in Madagascar that can do biomolecular work.
>
> (Ratsimbason, interview, 4/9/2013)

The Malaria lab here has also benefited from South–South collaboration and from the Panama ICBG. In Panama they do a *Plasmodium* resistance test using a fluorescence technique that they developed. Madagascan scientists learnt how to do this with ICBG funded training in Panama and brought this back with them. Dr Michel indicated that 'this is ideal for CNARP because the fluorescence is a more simple, inexpensive and useful technique. It is not problematic like radioactive analyses where you have to deal with the waste, storage and safety – many issues' (Ratsimbason, interview, 4/9/2013). All parts of the malaria laboratory were ICBG funded and this is important technology transfer for CNARP: 'the technology must fit the capacity, costs and maintenance. It is better to have the apparatus' they can use and maintain. The microplate fluorescence reader is reliable – we have had it since 2002 and it is still working and easy to maintain' (Ratsimbason, interview, 4/9/2013).

CNARP had benefited also from the provision of some basic equipment for a microbial research laboratory. The 3 main disease priorities in the country are malaria, diarrhoeal diseases and respiratory diseases. The microbial laboratory had been focusing on diarrhoeal diseases such as *Salmonella* and *Escherichia coli*, but had the capacity to study other infectious microbes (Ratsimbason, interview, 4/9/2013).

The chemistry laboratory had received more significant equipment including the HPLC and GC-MS machines noted above. The HPLC had been used for analysis towards ICBG research, and this laboratory had a number of posters explaining the identification of bio-active molecules with potential antiproliferative activity on cancer cells, which had been written into joint papers. The Gas Chromatograph was recently purchased for CNARP's use ($60,000) and is not for use in ICBG research. It has been useful for CNARPs own analyses of essential oils from plants for quality control prior to sale in national and international markets (Ratsimbason, interview, 4/9/2013).

At CNRE, Prof. Felicitee Rejo and Dr Rado Rasolomampianina were interviewed about their contribution to the ICBG through microbial and bacterial culturing, isolation and extractions, as well as research on microbial diversity from soil and marine samples. The 2 scientists stressed that they were very happy with the collaboration and that there had been many benefits for them and that extractions were still continuing for another 5–6 months to finalise the ICBG research. They indicated that they worked closely with CNARP and CNRO and so the monetary benefits felt there (e.g. the malaria research laboratory) was also of benefit to them for collaborative research. All departments and laboratories at CNRE had received some equipment and benefits from the ICBG. The 2 laboratories that had benefited most were the chemical and microbial analysis labs (Rejo and Rasolomampianina, interviews, 3/9/2013). Equipments included:

- Special freezers for drying of marine micro-organisms or other materials such as plant extracts prior to the extraction of DNA for analysis;
- Digital imaging equipment to image DNA from molecular extracts;

- Microbial DNA isolation kits;
- Thermo-vacuums for drying samples;
- Incubators for culturing micro-organisms from soil samples;
- Pumps and heating baths for evaporation, for drying marine algae, coral and sponges;
- Components for the maintenance of a gas chromatograph for analysing essential oils from plants.

As well as basic instruments, safety equipment and replacement parts for maintenance of some of the machines. Dr Rado noted that although much of this equipment was purchased to support ICBG extractions, it would have other benefits for CNRE in the future. He highlighted that collaborations with Prof. Russell Hill at Maryland would likely continue and that research on microbial diversity from soil and marine micro-organisms in diverse environments would continue with MBG (Rasolomampianina, interview, 3/9/2013).

Community upfront compensation funds

MBG have expended more than $216,085 in the last 5 years on upfront compensation projects in Ankafobe, Ibity, Oronjia, Pointe a Laree, Vohibe and Analalava. This figure includes carry-forward of funds from the previous 5 years of the ICBG. Expenditures of $117,428 by MBG on various projects have been allocated and spent from the current 5 year ICBG round (Kingston, pers. comm. 22/10/2013). Prof. Kingston estimated that CI had spent at least $100,000 over the past 5 years on community-based projects.

To implement community upfront compensation projects CI or MBG would hold meetings with the community to ask them about their needs and priorities for the funding. The country manager for CI noted that 'it was bottom up and came from them' (Andriamaro, interview, 2/9/2013). The community would put forward many propositions, but they helped them prioritise some to fit the funding level and so as to maximise beneficiaries (Andriamaro, interview, 2/9/2013).

For example, schools, public halls, wells and other items were often chosen that provided benefits to the majority of the community. The facilitators work out with the community the likely duration and budget for the project and estimate the costs of contractors (usually locals). These proposals are then developed to be put forward to the consortium for consideration. To secure approval, the community must commit to 'funding' 10 per cent of the project (to ensure ownership/commitment to the activity or resource). The participation of the community in the project can be in the form of materials (such as sand or stones collected locally that are used in construction projects) or labour. A local committee must be established to manage the project, and contracts with the contractors are provided. Final project documents are submitted to the MBG or CI site facilitator to obtain the signature of the Malagasy consortium members and to release funds into MBG/CI and community bank accounts.

Then work starts at the site and is monitored by the MBG or CI facilitator with regular reports to the Malagasy consortium members. These steps are intended to ensure transparency (interviews, Birkinshaw, 2/9/2013; Rahari-mampionona, 6/9/2013; Raharimampionona, 2012; Andriamaro, 2/9/2013).

For simplicity, upfront compensation for communities is discussed below in terms of the 3 stages of the ICBG. Although the Zahamena site was not visited during fieldwork for this book, a number of CI and MBG staff who had been involved in the projects there discussed it in interviews. Interviews were then also undertaken at some of the projects in Diana region and in the central region, which were project sites during the second and third phases of the ICBG.

The communities at Zahamena opted to use money from the ICBG upfront compensation funds to rehabilitate schools, grain silos for rice storage and to build a bridge for community access between 2 villages, and an ecotourist hiking trail. Kingston (2011) noted that the footbridge is used by 800–1000 people each year and the ecotourist trail by about 50–90 tourists and researchers. Due to the inaccessibility of the site and the several years that have passed since that phase of the ICBG, it was not visited as part of this case study.

Diana region – Oronjia peninsula

Near the city of Antsiranana in Diana region of Madagascar (the far northern tip of the country) are several sites that were part of the ICBG. The Oronjia peninsula and surrounding locations included sites of both terrestrial (plants initially and then soil microbes at the Oronjia forest and Montagne de Français) and marine bioprospection at Ambodivahibe (focusing primarily on marine microbes and algae – it was unclear if sponges and corals were collected). Bioprospecting began in 2006 and continued until 2008 in Oronjia forest by MBG and CNRE. Marine bioprospecting was conducted in 2010 by CNRO (Razafitsalama, interview 7/9/2013).

A number of interviews were conducted here on the 7th and 9th of September 2013. During the interviews no concerns were raised about the bioprospecting activities. The community seemed generally aware that some research had occurred but did not seem concerned – perhaps, because of the predominant focus of soil and marine microbes. Members of most villages noted that they did not know anything about the results of the research, and so we explained further about the progress of the research in the United States and Antananarivo.

Luciano Andriamaro from CI indicated that during the second phase the ICBG had built troughs for animals and 4 wells for villages in the area – water was raised as a major priority for the area and so the community chose to allocate funds primarily towards this. Some money had been allocated to a project relating to the rearing of chickens – however, a number of stakeholders had pointed out that this was not a very successful project because a disease infected the chickens and they all eventually died. In the third phase of the ICBG they had funded the construction and rehabilitation of schools in the area, toilets and sanitation, a common house for meetings and events. In 2010 they had also purchased boats and fishing equipment for at least one community. Very recently (late 2013) they had contributed 2 wooden motorboats for communities for fishing and to bring products to town. In addition, they purchased life jackets, GPS, binoculars and outboard motors. These boats will also be used for ecological monitoring and patrolling the marine protected areas for prohibited fishing activities (discussed further below) (Andriamaro, interview 2/9/2013).

At *Ankorikihely* (the village at the road junction leading into the Oronjia forest), 7 men were interviewed including the head of the fisherman's association, the head of the farmer's association and 5 other fishermen (several others attended but were silent). This project was selected to improve the productivity of the fisherman (a primary source of income and subsistence for the community). The ICBG funded a boat and a motor, a small building (fisherman's storage/cooperative and meeting place), life vests and nets in 2010 (Razafitsalama, interview, 7/9/2013).

The men indicated that they had chosen the boats because they want to keep working and to improve their activity – to have better equipment for fishing. They indicated that they would rather be fishing than to be in the forest collecting wood for charcoal like they had in the past (and which was being discouraged by NGOs, MBG and CI). After initially receiving the materials they were happy. However, the fisherman's association have had problems of maintenance and management of the boat and equipment. The problem of maintenance here was especially prominent in the dry season from May to November when conditions are very windy, and so the boat and equipment is often sitting idle and not being maintained. It was also during this period that the threat to the forest increased as a result (interview, 7/9/2013).

During interviews, the community complained that they had only 1 boat for several men and there were sometimes arguments over who could take it out on a given day. One boat was not enough for the whole association. They noted that they had no money to maintain the motor and make repairs to the boat. The men noted that only once had they received benefits from the ICBG and that they hoped they would come back and fund more. There was a misunderstanding between one of the men and the MBG staff while we spoke to them – the man had a spreadsheet of funds and expenditures and clearly expected that their community would receive other benefits. Jeremie then explained that the benefits had to be shared across several other villages in the region (interview, 7/9/2013).

It was noted that 2 wells had also been installed using the ICBG funds in a nearby school in Ankorikihely. These were quite full and were used by the teachers and students.

The community indicated that they had been recruited during bioprospecting and conservation fieldwork as guides for the MBG staff and also CNRE. All the staff involved were Malagasy – the US scientists were not involved in field collection in this region. The different Madagascan partners collected the samples, extracted materials and shipped the samples to the United States for further testing. The people in the United States reputedly knew only the code of the sample, not the specific name of the sample (interviews with Fishermen and Razafitsalama, 7/9/2013).

The community was not concerned about the research by MBG. However, they were interested to know more about what the US scientists in Maryland (and Virginia) had achieved through their research. They especially wanted to know more about the ecological condition of the fish – given the importance of the fish for their livelihoods (interview, 7/9/2013). Recently CNRE had been investigating incidents of poisoning from eating fish, sharks and turtles in this region, and was looking for external partners to assist with this research.

At *Ambodivahibe* village we met with the president of the village, the head of the parents and citizens association and 2–3 other men. This village had received a school as upfront compensation from the ICBG. Two of the interviewees responded that they were happy to have received this and that education has improved for the children in the village as a result. The following comments were made:

- 'Indicators [of improved education] are that every year every child is passing their exams. Here at the [primary] school since 2006 every student has succeeded.'
- 'Before the building was wooden and during the rainy season it was problematic (leaked) and also it is very windy here sometimes [the new school is brick/concrete].'
- 'There have been some problems. Here they need to pay the teachers. An association of parents has had to pay money for the teacher. Only one of the teachers is paid by the government but the other by the association of parents. They need 2 teachers because there are now 60 students.'
- 'We chose the school as compensation for collection in the forest and mountain – here in Madagascar there is no law on this but they are on the way to implementation of a law.'
- 'We have heard the results of marine and terrestrial research [by MBG and CI] but it is not enough. We are hopeful that we could hear more results. We have questions of access to information from researchers. We'd like to know more – for example, we don't know why the marine bioprospecting was important.'
- 'We are happy for the conservation of these reserves and we are happy to help maintain. We are happy for the gifts we get from the ICBG. For the Malagasy people, education is the best gift to improve our lives and economy in general.'

This community was also involved in forest and marine conservation activities which the ICBG has contributed funds to and which MBG and CI help facilitate. These are discussed further in the 'conservation projects' section below.

Ambavarano village was the next village where interviews were conducted on the Oronjia peninsula. This village had received a village hall, a boat (see

Figure 3.2 below), a digital camera, outboard motors, life jackets, nets and binoculars from ICBG upfront compensation funds. This was all received in 2013. In this village we interviewed 7 women (including the president of the village) and 2 men. The following comments were made:

- 'We selected these materials (e.g. the boat and motor) to improve our collection of fish and products from the sea. Now there has been low production because of bad management in the past. We now have a marine reserve and patrol a zone using this equipment to improve the resources. Fishermen come daily from a long way away so the community have to constantly reinforce the marine area.'
- 'Before we did not have a *dina*[3] and protected area, but now we do have. That is why we asked for a boat. Elaboration of the reserve began in 2008 but implementation of the *dina* only occurred very recently. The reserve was established in 2011 and markers delimited the area. All activities are forbidden in the area for 3 months (it is a temporary protected area).'
- 'We know the donations were for the research conducted near here. We don't know much about the process of the research. We don't have much information about the results yet – we know some information only. The researchers did seek permission – we think so. They sought permission to stay here for about a week or so when it was being conducted.'
- 'The building is for the storage of this equipment provided by the ICBG. The hall is also for meetings and receptions. It is very useful for us.'
- 'We also received a project for water troughs [several years ago] for the animals which we are happy with.'

Ambodimanga was the last village visited for interviews on the Oronjia peninsula. This is a small village so there was less funding spent and were less benefits here. We interviewed 2 women about the establishment of a well here in 2006:

- 'The well is very useful here. They have a problem though that people from the other village come here to take water during the dry season. This is not enough water for the two villages.'
- 'This well is safer – it has a lid so that kids are less likely to fall in. It is also higher walled.'
- 'In the early morning the other village comes with many cans – the other village should ask permission. A lock would help so that they can limit the use. They come and take 4 cans per person and the other village has many people. The other village has 20 wells up to 30 m deep, but there is no water! These were built by other NGOs.'
- 'Digging the well is important maintenance – the community takes care of this.'
- 'We don't know about the research. We don't know why we got the well. It might have been provided by a mayoral candidate.'
- 'We are not involved in conservation activities but want to be' [the women provide details to the MBG staff].

Figure 3.2 Fishing boat near Ambavarano village

Source: Robinson, 9/9/2014.

From the range of different projects funded by ICBG and implemented by MBG and CI in the Oronjia peninsula region of Diana, the communities seemed generally happy with the impacts that they had made. Notably many of the concerns and problems that they had were associated with inter and intra-community conflict, and due to problems of cost to maintain the materials or buildings.

In addition to these materials that were specifically discussed by the community, it was noted by MBG staff that there were several other additional projects that were ICBG funded. There were so many ICBG-funded small projects that we did not have time to interview people about all of them (and some were attempted several years ago with limited success – e.g. the poultry breeding project). These included:

- A vegetable growing project,
- Poultry breeding,
- A micro-dam,
- Another school (a total of 3 in the area),
- A total of 3 boats,
- Two other wells (4 in the area),
- Another 2 troughs.

Table 3.2 Benefits received as upfront compensation near Ibity

Village / commune (Fokontany)	Benefits
Sahamalola	3 bridges
Sahanivotry	1 school classroom
Manandona	Handicrafts shop, new roofs for several market buildings, upgraded platform and roof for meat section/butchery, upgraded toilets
Ibity	Upgrades to markets including a new fence, new gate, new sections of shops and new roofs for some shops.

Ibity and surrounds

Near the city of Antsirabe is the Ibity Massif, which is an MBG project site, a community-protected area and a soil bioprospecting site for the ICBG. MBG staff conducted sampling at this site with a local guide. As with the other sites, members of the community who were interviewed indicated that they had been informed about the soil sampling and had no concerns (interviews, 5/9/13). Several local villages in the Antsirabe II district of Vakinankaratra region were visited: Sahamalola (South of the Ibity Massif); Sahanivotry (near the National Road to the southeast of the Massif); Manandona (on the National Road east of the Massif); and in Ibity town/commune (North of the Massif, 10 km off the National Road). Table 3.2 outlines the benefits received in each community.

At *Sahamalola* 10 local men were interviewed – some individually and others in a group. This included the head of the Fokontany (village/commune), deputy head, 2 elders and several other men. All of the interviewees commented positively on the contribution that the bridges had made to the community (see Figure 3.3). To illustrate the following are quotes from these men:

- 'Thanks [to ICBG] for the bridge funding. It is very good for us. One of my family collects beans, maize and other local products. With the bridge he can access these with the truck. Otherwise he would have to carry it on his head for miles.'
- 'The bridge allows many tourists and foreigners like you to see the town and south end of the Massif and this is good for the community.'
- 'We are very pleased with the people who gave the money . . . we hope that we are both beneficiaries [donors and the community].'
- 'At night if we are unwell we can go to the hospital or clinic . . . Or pregnant women. We can take a car rather than walk. So now there is a greater flow of people in and out . . . locals, foreigners and visitors. . . if there are events we can get money from visitors.'
- 'For the village it is easier to take produce to the national road by car or truck and to access the market (e.g. at Manandona).'

- 'It is also important for security and is easier for police to get here . . . There is slash and burn here but we [the village] try to respect the government's rules about fire. We tell people to respect the rules. But there is a security problem, robbers etc [that use the fire as a deterrent] and we can request to police or military. We can't control the fires.'
- 'In development terms it is useful. There are lots of associations in the village and if we ask for donors/NGOs to come to install irrigation etc, they can access the village and build it.'

In addition to these positive comments about the impact the bridge has had on the community, the interviewees made several other comments. The head of the Fokontany noted that 6 villages and their committees were responsible for the maintenance of the road. He thought that it would be useful to have a roads committee to make rules and to take a small levy for road maintenance. Mamisoa, the local MBG facilitator, noted that the bridges and road were well maintained, suggesting that the community has taken a level of ownership over the 'benefit' they have received. One man noted that it would have been better if the bridge was entirely concrete because the wooden planks could be stolen (or burnt in a fire). In this location they didn't know about the bioprospecting as these individuals were not on the original committee and the collection activities were further north.

Figure 3.3 Bridge at Sahamalola

Source: Robinson, 5/9/2013.

At *Sahanivotry* 6 people were interviewed, including the mayor, deputy mayor, the principal and some members of the local conservation management committee/men employed to monitor and police the Ibity Massif protected area. The classroom built from the upfront compensation funds was used for a grade of middle school children (in the age band 11–15) and could accommodate 60 students (Figure 3.4). These individuals again commented positively on the impact of the upfront compensation they had received. This included the following comments:

- 'It was chosen by the local committee because it has an impact for the children. It is in the public interest. There was not enough space in the school so we needed new classrooms.'
- 'The benefit has been that we have increased the number of kids going to school. Before the commune and parents had a big problem, we needed more capacity.'
- 'This [classroom] has meant a reduced burden and higher attendance. This is also of benefit to the parents as they are not minding kids and can work' (interviews, 5/9/2013).

The maintenance of the building was provided by an association of parents and the commune which has a budget for the school, and to pay staff at the school

Figure 3.4 Classroom at Sahanivotry

Source: Robinson, 5/9/2013.

to clean it. They noted that the ICBG just funded the building itself, not the furniture inside. One of the interviewees indicated that they would welcome any further help or funds that can be provided as the furniture was quite old.

In this particular village some of the interviewees knew about the bioprospecting activities as they had been told by MBG in the past. They indicated that they had not been part of a formal agreement (but presumably they had been involved in the local committee discussions regarding the upfront compensation). They indicated that they did not know about the results of the research and asked if the US researchers could pass on information about their progress. When told by myself/my translator about the natural products research on cancer in the United States and the malaria research being conducted at CNARP, they were glad to know about it.

The men employed to monitor the Ibity Massif also discussed their role – as this is funded from ICBG 'conservation project' funding via MBG (a separate funding stream to the upfront conservation funds) it is discussed in the following section.

At the next town *Manandona* we verified the construction of a shop selling handicrafts, a stall selling seedlings, new aluminium sheet roofs for several market stalls and a number of other improvements to the local markets, all provided by ICBG upfront compensation funds. Here we interviewed a total of 10 people including the mayor, deputy mayor, head and deputy head of the local conservation committee, head of the local women's association, the shopkeeper at the handicrafts store, the mayor's assistant and 3 other members of the local committee. The local shop was stocked with locally made handicrafts including handmade silk scarves from silkworms collected in the adjacent mountains. Adjacent is the market selling a variety of clothes, fruits and vegetables, meat and other local products. The following relevant interview comments were noted:

- 'We are very happy the ICBG built this shop. We can sell our handicrafts here rather than travelling to Antsirabe. MBG have sent to us some visitors and this has generated further interest and money for the women. The local women's association produces silk and cotton clothes and tablecloths to sell here and so they benefit. One salesperson works here while the other women at home make an income from their weavings. We hope to find wider markets.'
- 'The seedling shop was only recently completed. It is better to be selling them here at the roadside than the old shop.'
- 'The new roofs are lighter (aluminium) rather than the old heavier tiled roofs which are causing the struts to sag. The tiles are also cracking and full of holes. The old one leaked in the rain. If they have a storm or cyclone the new roof is better.'
- 'They [the committee] chose the roofs because people were complaining about the old ones. Even now they still want to upgrade the old ones. In hot weather the tiles may crack.'

- 'The new roof is better because it is pitched and allows more air in. Customers can see more and more people can come inside. Also the shop-owners would have to finish mid-morning if it was going to rain. With the new roof they can stay longer and sell more things.'
- 'The sellers can better communicate with each other as they aren't hunched under the awnings. They used to be bunched together to avoid the holes and leaks.'
- 'There are more tourists and people stopping because it looks neater.'
- 'The meat section was upgraded with a new roof. It is more hygienic for meat and there is a new platform for cutting it. The toilets are also upgraded for sanitation and hygiene.'

These interviews indicate largely positive benefits and impacts from the upgrades to the markets. Some commented that they still needed more roofs changed as still about half the roofs were tiled and cracked. The mayor indicated that the commune had hired a guard to ensure the site was not damaged at night, and that there is also a budget for maintenance.

From this group there was one person from the local conservation committee who knew about the bioprospecting research – he was told by MBG about it. He claimed that he didn't know much about and was not really concerned about the soil sampling and transfers to America for research. Other comments focused on the conservation activities on the Massif.

The last site visited in this district was *Ibity* town/commune. Here, the ICBG has also improved various aspects of the local markets (Figure 3.5). A total of 6 men were interviewed at this location including the mayor, the president of the management committee for the Ibity Massif, the guard of the markets and 3 other men. The most relevant comments included:

- 'Monday is the market day here. People are very pleased with the upgrade. Now they are more happy and better arranged to sell their items.'
- 'The fence was installed so that sellers must enter through a big gate. They pay a fee to enter. This generates money for the mayor's budget and the commune, so this is very important. The money at the gate goes towards cleaning, maintaining and painting the markets.'
- 'The improvements to the markets are important because we have seen increasing numbers of tourists here. It is more attractive.'
- 'There are a number of extensions to the shops so more people can come and sell things, and new roofs that don't leak. There are new toilets from the gate takings [not directly, but indirectly a benefit from ICBG funding].'

In this group of interviewees, one man had been involved as a guide during bioprospecting activities. The MBG officer at the site, Mamisoa, had been involved in the collection of soil (bioprospecting) and plant identification activities (for herbarium vouchering only) at this site (and in Zahamena and Diana). He noted that there was no training really for local people – some of

Figure 3.5 Ibity markets with part of the Ibity Massif in the background

Source: Robinson, 5/9/2013.

them just received some basic pay for their work as guides or porters (as noted by Neimark, 2012). However, the training of Malagasy people who work for MBG was highlighted by a number of interviewees.

The remainder of the comments made by the community were about the conservation of the Ibity Massif site (discussed below).

Ankazobe

The following section examines a number of upfront compensation projects near the town of Ankazobe in Analamanga region in the centre of Madagascar. Here 2 sites were examined and several interviews were conducted at Firaranzana village and Ankafobe forest on 6 September 2013.

The village of Firaranzana is located on the top of a plateau, surrounded by a savannah-like landscape, some native forest remnants in valleys and some eucalyptus and pine plantations. The village surrounds were bioprospected by CNRE during the third phase of the ICBG and they received 2 classrooms, tables and classroom furniture, toilets and 4 wells as part of the upfront compensation. The community had requested the construction of a school in the centre of the village because the old school was damaged by a storm and its roof was leaking and walls were crumbling. The ICBG also contributed to repairs to the buildings. It was indicated that this was a major problem for teaching and that it also meant that books stored in the building would be water damaged if it rained. It was also noted that the teacher and his wife live in one of the

Figure 3.6 The school at Firaranzana with the distinctive mural painted on the side

Source: Robinson, 6/9/2013.

classrooms because they did not have a home otherwise. Sixty students were taught in the school. In addition, MBG painted a distinctive historical mural on the side of the school (see photo). The mural is part of an ecotourism project facilitated by MBG in the community, and the image links botanical and cultural diversity and heritage (Figure 3.6). It explains the different activities in the community in the past and today (e.g. cattle grazing, charcoal production). The mural and related ecotourism activities were also supported by ICBG (these are discussed in the conservation project section below).

At a meeting with many members of the community was the head of the Fokontany, the head of the parents and citizens association, the teacher and about 25 parents. There were many comments from the community including the following:

- 'At the beginning we only had the old building with holes in the roof, so we appreciate this. The results are only positive. The kids keep improving and learning.'
- 'Thanks very much for the funders on behalf of the parents. We are very happy to have received this.'
- 'We are pleased this is built because otherwise we would have to send children to Ankazobe town and pay fees and daily transport. The distance is too far and the cost would be 2,500 Ariary per day – this is too much for us. So it is great to have this new school.'

- 'Thanks – we are very happy to have this school. It decorates our village [refers to a mural painted on the exterior wall of the school]. It is iconic in this area. It is also good for ICBG to do charity.'
- 'We are very happy with the wells. We used to have to walk about 500 m or more to get water. Now it is in the middle of town . . . we were suffering and these new wells relieved the suffering.'
- 'We are very happy with the wells. It is not enough – there is not enough water for the whole village, but we are still very happy.'
- 'We are happy to have time to work and to not have to mind the kids. The children are coming every day of term and learning lots.'
- 'Last year and this year we can compare the results of the children in the CEP (entry to middle school exam). Four students went to take the exam and all have passed. This year 7 tried and 5 have passed. This is an improvement on the past when only a few had taken the exam and passed.'
- 'We recognise the efforts of the teachers. We are not able to pay them enough; we are harvesting the fields so we can pay them with rice. We feel bad about this. But the building has helped.'
- 'Before the school the children were quite ignorant of many things. The rate of children dying because they were playing or running onto the main road was quite high. They did not know to even look for cars. They have to be schooled and learn basic things – also to learn literacy. It is also important to keep them busy. So the community decided as a whole.'
- 'The school was the top priority because the building was in disrepair and it is too far to Ankazobe. The wells were second priority.'

In addition to these comments it was evident that the community had taken ownership and responsibility over the new asset. The parents association arranged regular cleaning and agreed there would be painting of the classrooms every 2 years. It was indicated that 80 per cent of the money for maintenance came from parents while the rest came from the commune.

Although the wells had been gladly received, members of the community also noted that lots of people were using them to fetch water and so they quickly dried out. The community was planning to dig out more wells in the future.

Conservation projects

In addition to this, a separate pool of ICBG funding provided by the FIC of the NIH is used to contribute to local conservation projects. An estimated $30,000 per annum is provided by the ICBG to MBG (and a similar amount is provided to CI) as co-financing for a number of projects at or near the bioprospecting sites. Notably, MBG Madagascar has 11 conservation projects which have all benefited in some way from the ICBG. Despite the distinctly US name 'MBG', the MBG office in Madagascar is a relatively autonomous program of activities (the largest foreign program of MBG) with a relatively limited connection to St Louis, and acting much like an

NGO to facilitate conservation and community development activities across the country (Birkinshaw, interview, 2/9/2013). The MBG Madagascar program has been in place for 25 years and now employs more than 150 staff members, almost all of which are Malagasy – Chris Birkinshaw is the one current exception (MBG website, 1/11/2013; interview, 6/9/2013). Once they receive the funds from the United States, the MBG office in Madagascar is not restricted in how they use them. They receive funding from multiple sources for these 11 project sites and so the ICBG conservation funds have often been a useful way to bridge gaps in funding for useful activities. This includes the following:

- Fire breaks,
- Community policing,
- Local management committees,
- Control of invasive species,
- Vegetable gardens,
- For raising community awareness about conservation or fires,
- For the purchase of computers, tents and GPS,
- For the payment of salaries and wages of MBG staff or contractors working for MBG,
- For the procurement of seedlings and equipment for propagating native species (Birkinshaw, interview, 6/9/2013).

CI has also benefited from ICBG funding for conservation activities. This has primarily related to marine protected areas, although CI also received funding for terrestrial conservation activities in Zahamena in the earliest phase of the project, and had some involvement in terrestrial activities at Oronjia (though these appear to have been largely been managed by MBG in recent years) (Andriamaro, interview, 2/9/2013).

Conservation International indicated that they had assisted with the establishment and enforcement of marine protected areas in bays on the east coast of the Oronjia peninsula. These marine areas do not have definitive legal status but are currently under temporary protection as recognised by the government, or community protection. A process is underway for the formal listing of these areas. CI have helped create 8 marine reserves in the area (4 permanent and 4 temporary). There are established local rules called 'dina' (interviewees in the communities noted that these were in place in some of these locations before the entry of CI to these areas for the ICBG projects). There are 4 marine area associations – 1 per village – and they manage the reserves (no more than 3 each). Within these areas there are permanent no take zones. There are also temporary marine reserves where the dina of the local community should be respected. These temporary marine reserve areas are closed for about 3 months and then re-opened for fishing. All of these areas have to be patrolled by the associations, especially the permanent protected areas, to prevent poaching by other communities and their fisherman from along the coastline (Andriamaro, interview, 2/9/2013).

It was noted by CI that they had seen many benefits for conservation from the ICBG funding and that the communities had been largely happy to be involved, with very few complaints. There was now a funding gap whereby CI was having to look for other funders and opportunities to continue community conservation and livelihoods activities (Andriamaro, interview, 2/9/2013).

Interviews were conducted in the same communities as the upfront compensation projects to also determine the impact of ICBG assistance for projects facilitated by MBG and CI. While there was often co-funding for these projects, both CI and MBG staff indicated the usefulness of the ICBG funding to cover activities that might otherwise have been neglected, or as bridging funding.

Diana region – Oronjia peninsula

The community at *Ankorikihely* had been encouraged away from wood collection for charcoal in the forest through the purchase of a boat – this has primarily been successful in the wet season, but becomes problematic during the dry season when winds increase and navigation for fishing is more challenging and dangerous, so they are not able to travel far and this affects their catch. In Ankorikihely, they participate in protection and patrolling of the forest and in plantation of trees to restore some areas. It was noted that this is quite challenging because livestock seek out and eat the new plants. This first village was not involved with marine protected areas as they are located on the inside of the peninsula near Ramena (interview, 7/9/2013).

The forest is both a community protected area and is recognised by the government as a protected area. The size of this protected area is approximately 1,648 ha. The community noted that much of the responsibility is on them for the management of the forest. The main threats are charcoal making, slash and burn agriculture, the pasturing of livestock, collecting timber and some hunting. It was estimated that 90 per cent of the community was aware of the conservation challenges and need to protect the forest, but the main problem is poverty – the people need to do something to cover their daily needs. The options are fairly limited (especially since the political crisis and a decline in tourism) to fishing, charcoal production, livestock herding, small scale agricultural production, basic hunting and gathering. The members of the community receive compensation for patrolling the forest. Jeremie from MBG estimated that about one-third of the MBG funding for the area is provided to the community, and then much of the rest is paid to MBG staff salaries (and these local staff are also often from the community). For activities such as patrols they pay 1 person 5,000–8,000 ($2.20 –$3.60) per day and Jeremie indicated that the community seemed happy with the pay as a supplement to other daily subsistence or economic activities (daily incomes are often less than $2 per day in many parts of Madagascar). There are patrols every week and 10–15 people are included in each monthly roster. Part of this income is derived from the ICBG conservation project funding (Razafitsalama, interview, 7/9/2013).

When the community was asked if the conservation activities were successful, they said yes but there are fairly slow improvements since 2010 when MBG have been involved. They noted that there has been a decline in the number of people conducting destructive activities in the forest in recent years. They also recognised the importance of the forest for rainfall and water retention in the soil. The fisherman had not known much about the presence of a threatened species *Delonixvelutina* (listed as critically endangered in the IUCN Red List) in the forest and had been using it for the production of their boats – they were now aware about this and have restricted use.

At *Ambodivahibe* the interviewees from the community also had many comments about the conservation of the forest and marine areas. They discussed the nearby Montagne de Français (Figure 3.7), Ampiue forest and Ambodivahibe marine protected areas:

- 'We felt the importance of biodiversity even before the ICBG came. We had marine and terrestrial rules in *dina*. This protection existed before the organisations came. The rules are only for traditional use, not exclusive use. Trees can only be used for house and boat construction.'
- 'There are also strong rules for marine protection. The problem is the delimitation of the area and conflicts because of this (both marine and terrestrial). But now the resources are strongly protected.'
- 'The unclear delimitation of the forest from the village in the west (Ivovona) has meant that they have had problems with the community in Ambodivahibe. If the delimitation is not clear, the people from that village will destroy the resources. They have no *dina* there. For example, in marine areas, these people use destructive techniques on the coral – they use large nets. Here we use small nets.'
- 'We are funded for patrols of the marine area by CI [seem happy with funding]. The level of marine and fish resources are the same as before the ICBG and despite some conflicts [with other fishermen/communities].'
- 'The forest is a sacred area for us and contains burial sites, so we have been protecting it anyway. We are happy with the assistance for conservation of the areas.'

In Ambodivahibe the interviewees wanted to emphasise that they were already practicing conservation – they had a conservation ethic already and this was related to the need to conserve resources as well as to respect sacred sites. The community members indicated that they were generally happy with the additional assistance they had received from the partner organisations.

At *Ambaravano* the community made several comments about the temporary marine protected area that they now patrol:

Figure 3.7 Montagne de Français protected area

Source: Robinson, 7/9/2013.

- 'As a girl [approx. 40 years ago] there were many fish – lots to catch. Since then there has been a gradual decline. Things seem to be improving lately.'
- 'We feel a big improvement. In 2012 we did not have good harvest because of the lack of equipment. After patrolling we think there has been a big change – after the fishing season closed they were able to collect 2 tonnes of octopus. After another year it will increase even more.'
- 'There are many challenges because this bay here is one of the best in the world in terms of biodiversity. CI has many activities and even bring people from the community to Tulear (in southwest Madagascar) to learn.'
- 'All people here have agreed to implement the marine protected area. The main problem is with people from the other villages coming. People from Ambolobozikely come from the south.'
- 'The *dina* is officialised to the court now – if it was not official then it would not be effective. If we catch someone we can apply the *dina* directly to them. If caught they have to pay a fine to the president of the fishermen or president of the Fokontany. We use this money 50–50 for the fishermen and the community/Fokontany. We have already caught many people. Not all people doing the infractions can be caught. Sometimes they refuse to accept the provisions of the *dina* so the community calls the Gendarme . . . If the marine reserve is well managed then the people from the community will have enough production.'

- 'CI did much training especially on the management of reserves and breed-ing cycles, and for management of their organisations. We are happy and even want more training.'

This community seemed in contrast with the other village nearby (Ambo-divahibe) with regards to their lack of conservation (now primarily focused on marine resources) prior to the activities initiated by CI and MBG. As a result, the community strongly emphasised the impact that CI (and MBG) had made on their activities and marine resources, with the expectation of ongoing improvements by continuing to manage the reserves. A key question will be if the community can and will sustain these patrols and conservation activities into the future if CI and MBG do not have the resources to continue their work there and to continue paying for patrols.

Ibity and surrounds

Although *Sahamalula*, the southern-most village interviewed was a beneficiary of ICBG upfront compensation, they were not involved in conservation of the Ibity Massif (they are located a few kilometres south of the Massif). They did not have their own *dina*, but instead abided by government rules restricting the use of fire as an agricultural tool. They noted that the bridges provided by the ICBG had assisted with their security in the area, benefiting conservation indi-rectly. This allowed gendarme to enter the area if there were people lighting fires in the forest and grasslands as a distraction to conduct robberies.

Sahanivotry commune had a conservation committee for the protection of the Ibity Massif. The interviewees indicated that their role was to educate the public about the significance of the massive as a site of unique geological, biodiversity and cultural interest. They communicated to people about the prevention of damage to the natural ecosystems as well as monitoring and reporting to the authorities. They tell people about the disadvantages of irresponsible use of fire as an agricultural tool or for creating new growth in old pastureland, its impact on rice fields, soil and rainfall (evaporation).

The men noted that security is a big problem for protection. The Massif is very large and wide and so is hard to patrol. Thieves often hide in the Mas-sif. They use fire to cover their tracks or as a distraction while they steal zebu or they steal from people's houses. Wild fires have decreased as a result of the patrols but have not stopped. The people who live near the Massif are quite convinced about the need for conservation and to avoid lighting fires but the thieves are the problem (interviews, 5/9/2013).

The men are paid by MBG in 3 villages (1 group each), and the ICBG conservation funds have contributed to this. They have 10 days of payment to conduct patrols for each village per month. This is usually 50,000 Ariary ($22.40) per month for 5 people ($4.50 each). They have received t-shirts from MBG so as to be recognisable as patrolmen and flip-flops. The men noted that communications are a big challenge – they generally run downhill or call for help to nearby people. If the fire is a long way away they have to rely on people. They indicated that mobile phones would be helpful and also fire breaks and water to put out the fires (they sometimes have to beat at the fires). A link between the 3 villages, such as a bridge across the river and road would be helpful (interviews, 5/9/2013).

In *Manandona* the conservation committee made similar comments, except that they were paying more men (a total of 10) to do the patrols and that each man was paid 5,000 Ariary ($2.25 each). These men have specific zones that they patrol, usually near their property/village. They also have a sign in book to ensure they conduct patrols to receive their payment from MBG via the president of the committee. This village has not yet received its t-shirts and flip-flops for the men, and also requested phones or radios. They were using whistles to alert people of fires. Mamisoa from MBG estimated that at least 50 per cent of the fires were started by thieves/criminals (interview, 5/9/2013).

In *Ibity commune/village* there were again similar comments and issues raised. They noted that tourism had increased since MBG and CI had been involved and had put up signage, and helped create tourist interest in the site for hiking and climbing. Some Malagasy tour operators had been bringing tourists here on some Mondays, and this provides some local economic benefits. When asked how MBG had helped, the patrolmen noted that they had received lots of support in terms of information, the allowance, shoes and shirts, posters, meetings (including some money for this) and brochures. This community also asked about how communications could be improved for the patrols. They also noted that one man had helped with the bioprospecting in 2011, although it was acknowledged that no real local training was provided in terms of botanical inventory. Rather Mamisoa – the local Malagasy MBG staff member (based in Manandona) had been employed to identify plants on site (as well as in Diana and Zahamena) and had received some training and several years of part time employment in relation to the ICBG activities (interviews, 5/9/2013).

It seems like the support MBG has provided, although minimal in terms of monetary expense (of which the ICBG contribution is a fraction), has had some impact towards limiting fires started by local communities. In their annual report (MBG Madagascar, 2012), they noted that fire in this region is one of the most problematic challenges they face in their conservation projects. This is severely hampered by the security situation and several local community interviewees noted that there was nothing they could do at night-time when fires were lit (thieves were generally blamed) because they had to protect their homes and families. Apart from these challenges, there are a number of other monetary and non-monetary benefits that were acknowledged as arising from this work (e.g. income, awareness and capacity-building and tourism benefits).

Ankafobe

A small (6 km²) community managed forest has been supported at Ankafobe (near Firaranzana village in Ankazobe district) by MBG, including with ICBG conservation funds. A voluntary local group (*Vondron' olona Ifotony*– VOI) was established by the community in 2005 and patrols and protects a valley forest remnant in a landscape that has been largely cleared and burnt. The forest remnant is significant because it contains a critically endangered species of tree – Sohisika (*Schizolaena tampoketsana*) – of which there are only an estimated 200 known plants in existence. Five other locally endemic plant species are found here in valleys and 4 lemur species (one common species and some less common). Hundred plant species are found in the valley, including a number of medicinal plants. To protect the valley from fires that have burnt up nearby hills in the past, fire-breaks (prescribed burns) have been burnt into the hilltops with permission from the Forests Department. These have cost about $2,000 and have been funded by ICBG in the past. Apart from the important fire-breaks (Figure 3.8), VOI and MBG activities at the site (all funded considerably by ICBG) include the following:

- Establishment of a campground/picnic area and use of these sites for school camps to educate children about the forest,
- Establishment of a local nursery (5,000 plants per year) for replanting up the hills to expand the forest and prevent erosion,
- Contracts for the head of the VOI, a nursery carer, and a patrolman paid by MBG,
- Ecological research, restoration activities and monitoring of microclimatic conditions,
- Development of ecotourism and the maintenance of viewing platforms, 2 km of trails, signs and bridges (platform and trails funded entirely by ICBG conservation funds),
- Fish ponds and bee-keeping as sustainable livelihoods activities.

Jeannie – the MBG coordinator for the site – indicated that almost the entire $5,000 budget for conservation and tourism on the site has been provided by the ICBG (approximately 90 per cent). Additional co-financing had been received from private donors for research and from the GEF Biodiversity Small Grants Programme for a multipurpose building at the entrance to the site.

The members of the VOI we interviewed indicated that they were very happy with the assistance and funding they had received. They indicated they had had 100 per cent success here since they started in 2006 in keeping fires out and maintaining the forest size. They were happy to have received wages for the construction of the trails from MBG. They were also expecting tourist numbers to now improve with some new signage – there had been improvements in the last year. They will also start charging people 1,000 Ariary ($0.45) for locals and 10,000 for foreigners ($4.50) to use the trails and campground (interviews, 6/9/2013).

This site appears to have benefited significantly from the ICBG conservation funding and support of MBG. These impacts have been felt not only for conservation, but also in terms of local livelihoods for several people.

Figure 3.8 Fire breaks across the hillsides (left) to protect the valley forest at Ankafobe (right)

Source: Robinson, 6/9/2013.

Potential monetary benefits

A formal Research Agreement was negotiated among all the parties to the Madagascar ICBG. This agreement included specific provisions for the sharing of monetary benefits to Madagascar, incorporating both the Upfront Compensation that supported the work described in this chapter and milestone payments and royalties on any drugs or agrochemicals discovered from a Madagascar plant, marine or microbial source. The milestone payments differed for each company. For example, those for DAS were in the 6 figure range. In some of the agreements, payments were related to specific milestones tied to the drug approval process. The royalty rates also varied by company, but both were based on a percentage of net sales of the licensed product, with the actual percentage being in the single digit range as is typical for early stage discoveries (Kingston, pers. comm. 20/1/2014).

As an example of the royalty structure, a new drug or agrochemical discovered at Virginia Tech would be licensed by Virginia Tech to DAS, depending on the nature of the discovery, and Madagascar would receive a portion of the royalty income received by Virginia Tech (Kingston, pers. comm. 20/1/2014).

The Madagascar portion was set at 50 per cent of the total royalties and upfront compensation payments for a drug or agrochemical that was an actual natural product. If the drug or commercial product was based on a synthetic compound based on a Madagascar natural product model but was significantly modified to improve potency, bioavailability and so on, then Virginia Tech would have contributed more to the discovery than just the isolation of an active compound. In this case the Madagascar portion would be reduced from 50 per cent to 20–40 per cent, depending on the extent of the changes and the intellectual property contribution made by Virginia Tech. Similar arrangements were built into the Research Agreement for discoveries made by other partners (Kingston, pers. comm. 20/1/2014).

Non-monetary benefit-sharing

There are a number of non-monetary benefits arising from the different aspects of the project. These include papers (discussed in the section on R&D), training in-country and through exchanges, provision of equipment (discussed in monetary benefits), employment and benefits for conservation. While some benefits can be considered incidental (e.g. employment on the ICBG), many others are direct and deliberate.

A number of Madagascan interviewees indicated that in-country training as well as workshops and activities make up an important part of the non-monetary benefits. Prof. Kingston has provided a detailed breakdown of training provided by the ICBG to host-country trainees (pers. comm. June 2013). The vast majority of these are Malagasy with only a few exceptions (e.g. training involving Chris Birkinshaw from MBG Madagascar), and they include training for both scientists and local people, typically run by CI or MBG.

In terms of long-term training (4 weeks or more) for host-country (Madagascan) residents, there are 22 people who have received training supported by the ICBG. Many of these were at the US partner organisations like MBG St-Louis (specimen collection and identification), at DAS (microbiology) or at Virginia Polytechnic Institute and State University (natural products chemistry). Some also received training in fluorescent antimalarial assays at the ICBG Panama. Others were also trained in botanical specimen identification at the French National Museum of Natural History (Kingston, pers. comm. June 2013). These latter 2 were confirmed in interviews with the recipients of the training (interviews, 9/9/2013). Often the training had a direct benefit towards the ICBG. For example Leontine Rahelinirina (CNARP) was trained at MBG in St Louis (USA) in 2001 in horticultural techniques for management of the CNARP medicinal garden (WLBC/MBG, accessed 17/10/2013). Despite this, much of the training is likely to have important future applications for both research and conservation activities in Madagascar as evident from several of the interviews conducted. For example, Jeremie Razafitsalama was sent to Missouri for training in botany which he uses as the regional manager for MBG conservation and livelihoods projects in Diana. In another case, a Malagasy PhD student who had some involvement in the ICBG research conducted his degree between the University of Antananarivo and the Saint Louis University (Razafitsalama, interview, 9/9/2013).

Host country people also received training which tallied to more than 1,329 individual people/incidences (of this it should be noted that some people may

have received training more than once) in short-term (less than 4 weeks) activities. These were typically facilitated by MBG or CI in-country and were often
employed alongside ICBG monetary expenditures for upfront compensation
projects or conservation projects. In addition, there are an extra 55 groups of
unrecorded size, from family units to whole villages or village associations,
which have also received training. These are often workshops or training ses-
sions for a matter of days or weeks. They include community livelihoods training
activities such as poultry rearing, market gardening, beekeeping, as well as conservation-oriented activities such as forest conservation workshops, training for
forest police and workshopping protocols for sustainable lumber exploitation.
On the other hand, there is considerable research training including botanical
and ethnobotanical field methods, as well as laboratory-based training on use of
equipment such as HPLC isolation methods (CNARP), and extraction of DNA
from soil and marine micro-organisms (CNRE) (Kingston, pers. comm. 2013;
Cao and Kingston, 2009; interviews with MBG staff, September 2013).

There were also other non-monetary benefits for laboratories such as
CNARP and CNRE. For example, CNARP had been involved in collecting voucher specimens of plants and were able to collect replicates which are
kept in a herbarium as a reference collection (where the microscopes and some
cabinets were also partially funded by ICBG). They were able to collect these
samples from the first 2 phases of the ICBG (Zahamena and Diana) (Ratsimbason, interview, 4/9/2013).

Impact

There were direct benefits related to the botanical inventory research con-
ducted alongside the field bioprospecting activities. From the fieldwork at
Zahamena, MBG were able to inventory 930 species from 128 families.
Some of this information has been published in an *Inventory of the Ferns of
Zahamena* (in French) which will benefit researchers, tourists and reserve
managers (WLBC/MBG, accessed 17/10/2013). Botanical inventories at the
Montagne des Français recorded 215 species of higher plants, 5 primates, 12
small mammals, 56 bird species, 40 reptile species and 19 amphibian species,
including the identification of some critically endangered species (Kingston,
2011). It was also evident from the interviews in Ibity, Oronjia and Ankafobe
that there were many conservation benefits arising from the vouchering and
identification of new species and new conservation priorities, as well as from
the direct ICBG support for conservation projects. Areas at all 3 of these
sites have received some form of community or temporary protected area
status with formal recognition pending government ratification in a number
of cases (i.e. Ibity Massif protected area, Montagne des Français temporary
protected area, Oronjia forest community protected area, the Ambodivahibe
Bay marine reserve, Ankafobe community protected forest). This has been
supported by MBG and CI through the collection of data and monitoring of rare and endangered species, as well as through community patrols

and monitoring. Related to the establishment of these protected areas, the potential for tourism has also increased at all of the sites visited. For example, Ambodivahibe Bay is a specular location near to an existing tourist destination with tourist lodges – Les Trois Baies. Given the relative poverty in all of these communities, economic development opportunities such as tourism could provide important income supplements for some people.

While it is difficult to gauge the 'success' of these conservation initiatives, there are certain indicators which can provide some guide. At Ankafobe the community was obviously strongly behind the conservation of the forest remnants and for several years had defended the area from fires, with some replanting of trees on the slopes (although this had been hampered by the hard laterite soils), retention and propagation of the Sohiska tree and other locally endemic species. Anecdotal evidence from the communities in Oronjia about marine protection and marine production suggests that temporary protection has been relatively successful in the short term. Communities also seem committed to protection of the forest at Oronjia, even though they are having to compete with other villages who have been entering the area to collect timber for charcoal and other purposes (MBG has been monitoring a number of indicators from 2012). At the Ibity Massif, MBG and community monitoring indicated that 2,808 ha had been burnt in 2012 and this is considered to be a much larger area than natural burn conditions in the hilly grasslands/shrublands in the area (MBG Madagascar, 2012). On a more positive note at Ibity, MBG is getting close to designating the Ibity Massif as a government-recognised protected area, having been allocated a *cahier des charges* by the ONE, listing expected conditions for the management of the site. The final stage of consultations with local stakeholders was being conducted in 2013 (MBG Madagascar, 2012).

MBG staff noted that infrastructure projects were generally considered to be more successful than some of the other projects such as vegetable growing and chicken rearing, despite these being specifically selected by locals (Birkinshaw, 2/9/2013). The MBG staff tried to emphasise projects which would maximise the number of beneficiaries – and so schools and wells were often chosen and built. For infrastructure projects such as the schools and wells, while generally perceived as successful, maintenance and sustainability will be a key issue over time (Birkinshaw, 2/9/2013). For example, wells would have to be re-dug or classroom roofs might need repairs after storms and this might have to be paid for and/or done by the community. Other projects such as the boats and fishing equipment for the communities at Oronjia had been successful in some instances (those projects in the early stages of delivery) while in one village there was evidently some frustration about maintenance of equipment and the inadequacy of the one boat for many men. Yet some indicators, like the number of students passing their exams, show a clear improvement over time for the community.

For the 2 government research centres (CNRE and CNARP) where staff were interviewed, the scientists were very pleased with the training they had received and the equipment purchased with the upfront compensation funding. The contact for ICBG from CNRO was not interviewed as he was apparently at a distant regional office at Vangaindrano in SE Madagascar, which was itself partially funded by the ICBG upfront compensation funds. From discussions with these scientists and other interviewees there was some suggestion that there could have been greater support for research on tropical diseases and diseases of high priority for Madagascar (by shifting the dominant research focus of the project on cancer somewhat). Notably, however, there had been substantial support for some laboratories to conduct this sort of research and analysis at CNARP. Indeed the scientists at both institutions lamented the end of the ICBG and were now seeking other external partners to continue research with.

To summarise the range of impacts we can use World Bank development and poverty indicators, which are often applied to integrated conservation and development projects (ICDPs). Table 3.3 summarises how the ICBG contributes to different types of 'capital' in Madagascar.

Continued monitoring of these impacts into the future will provide definitive evidence of the overall impact of the project for conservation benefits as well as for other aspects of the project.

Discussion

An interview with the ABS national focal point – Dr Naritiana Rakotoniaina from SAGE, revealed that there was still a need to clarify and improve the legal arrangements for permitting ABS in Madagascar. While SAGE were aware of

Table 3.3 Assets and impacts for Madagascar

Type of 'capital'	Impact
Natural	– Identification and conservation of rare and endangered species – Protected areas being established at a number of field sites and management plans being implemented
Human	– Training of Malagasy scientists – Training of Malagasy staff at MBG and CI – Training of local communities as part of ICBG projects – Joint publications benefiting Malagasy scientists – Improvement in educational outcomes in local schools – Potential drug development
Financial	– Upfront monetary benefits – Potential monetary benefit from milestone and royalty payments
Social	– Livelihood benefits from a number of upfront compensation projects and conservation projects/training – Established collaborations within communities and with MBG and CI for conservation and livelihoods goals – Scientific collaborations
Physical	– Equipment for the National Centres – Equipment for local communities (e.g. boats, motors) – Schools, wells and infrastructure for communities

the benefits provided to the national centres and also in some of the regions, commenting positively on those, they had some other questions about the ICBG project as a whole. Specifically, they were not very aware of the use of the results after the project, what would happen to the results and the extracts from the genetic resources,[4] and how these would be monitored now that they had left the country (also noted by Neimark, 2012). People had asked SAGE and the ICBG if transfers of the extracts and R&D to third parties had been made and there was some concern if this was restricted in the initial agreements made by the consortium. Would these third party transfers dilute the likelihood of eventual milestone and royalty payments? Those who are outside the consortium have never seen the contract and so they may have many questions about the future of the R&D (interview, 2/9/2013). Luciano Andriamaro, the Director of Science and Knowledge at CI, and the person charged with administering CIs component of the ICBG, further discussed the ABS context in Madagascar. She noted that it would be useful to have an ABS law in Madagascar to encourage future ABS projects which took a similar approach to this ICBG and also to ensure those future projects were transparent and validated (interview, 2/9/2013).

In local communities when people were questioned about the knowledge of the research, particularly the purpose of the bioprospecting, there was generally quite mixed knowledge. It was evident from my prompting in the interviews that some people were curious about the results, and some were even aware that the research had the potential to treat diseases like cancer. But others had little or no knowledge about why samples were collected. This highlights one of the challenges with bioprospecting research, especially for drug discovery, that the R&D timeline is exceptionally long and by the time any drugs are discovered, the initial field activities may be long forgotten (indeed, members of the community may have died or moved away). This also means that the external or foreign partners may lose communications with the original providers. In this case this seems to have happened somewhat about the drug discovery R&D, but not with the more pertinent research on conservation in the local areas. It was evident from the field visits to villages that the communities were in very regular contact with MBG and CI regional project staff and were well aware of conservation concerns and needs for each area. Luciano Andriamaro, noted that CI had summarised and simplified their conservation-oriented study results and put them onto posters in Malagasy language to be simple and easy to understand (interview, 2/9/2013). Similarly, MBG had assisted with detailed signage about specific species and conservation priorities at sites such as Ankafobe forest. However, a recurring theme in local interviews was concern about ongoing maintenance and funding for activities. Many lamented the end of funding for local activities in all of the project sites. On the other hand, the 15 years of funding in 5 year phases is relatively long compared to other grant schemes.

One of the interviewees noted that the long delays following leads in the United States were frustrating. They suggested that they should share more of their results in cases where they do *not* want to pursue a lead so that perhaps scientists in Madagascar can. After discussing some of the activities of the ICBG Papua New Guinea, the interviewee agreed it could be useful to do more basic chemical validation of traditional medicines that might be of use in Madagascar. Another interviewee noted that it was odd that the companies involved in the ICBG were not seen to publicly promote their activities as part of the collaboration, including their contributions to local communities and conservation. Although they are part of the consortium, to the public they are silent partners (interviews, September 2013).

Neimark (2012) indicated that in a number of interviews with scientists and collaborators on the ICBG that there was some resentment that much of the R&D was conducted in the United States and only basic extractions were conducted in Madagascar. Although my interviews did not reveal this same sense of discontentment, there was some equipment in the laboratories that were clearly for ICBG extraction purposes only and also training received was often directly related to the ICBG research. However, the scientists at both CNARP and CNRE pointed out much additional equipment that they had used for their own research. They also pointed out that much of the training had equipped them with additional useful skills which could be applied to other research (interviews, September 2013). I suspect that many of the earlier issues that may have arisen in the project were gradually improved as the project team learned and adapted over the 15 years.

Neimark (2012) also raised some concerns about the utilisation of information and samples collected for herbariums and listed in the TROPICOS database. Staff at MBG noted that they had agreements with the botanic gardens and herbariums that they shared voucher specimens which limited uses to basic scientific activity, and that they are dried voucher specimens – not easily used for extractions of biochemical compounds. They also noted that conservation activities by MBG, CI and also other NGOs and the government were and could benefit immensely from the information collected in the field and inventoried in the database. Indeed, the nomination of protected areas had benefited from the identification of endangered species and areas of ecological significance during fieldwork that had been funded by the ICBG. In addition, the training of local communities as well as training and employment of Malagasy staff at both MBG and CI was of significance for both conservation and local economic development. In general, the consortium stakeholders, local communities and even non-consortium interviewee spoke very positively of the ICBG, even while noting some of the challenges and concerns they had encountered. Unlike many ABS project which never yield real benefits for conservation or local communities, this project had contributed many benefits upfront (and potentially into the future), as well as potential enduring or multiplier benefits (e.g. additional benefits from training, or additional benefits that come from tourism at the community project sites).

After 15 years of activity in the country, and at the end of the project, there were many comments emphasising the 'partnership' nature of this project.

Compared with other ABS projects (further comparisons will be made in Chapter 11) the upfront benefits, number of local beneficiary projects, joint-publications, training benefits, equipment and conservation activities are extensive. In my opinion, this was one of, if not *the* most comprehensive and impressive ABS agreements made to date.

Endnotes

1 Initially the ICBG was funded by the US Agency for International Development (USAID) and this changed to USDOA.
2 Notably some are written by David Kingston as review papers relating to natural products drug discovery, or are papers by other non-Malagasy authors on taxonomic description of new species.
3 Dina refers to customary rules or laws held and enforced by local Malagasy people. It is a type of social contract, which may also be formally legally recognised.
4 It is worth emphasising that the ICBG consortium in Madagascar was not exporting genetic resources, but rather they were exporting biochemical extracts of the genetic resources.

References

Cao, S.G. and Kingston, D. (2009) 'Biodiversity conservation and drug discovery: Can they be combined? The Suriname and Madagascar experiences'. *Pharmaceutical Biology,* 47(8), 809–823.

Cao, S.G., Ranarivelo, L., Ratsimbason, M., Randrianasolo, S., Ratovoson, F., Andrian-jafy, M. and Kingston, D.G.I. (2006) 'Antimalarial activity of compounds from *Sloanear-hodantha* (Baker) Capuron var. *rhodantha* from the Madagascar rain forest'. *Planta Medica,* 72(15), 1438–1439.

Fogarty International Centre (FIC) website. '*International Cooperative Biodiversity Groups Program*'. Available at: http://www.fic.nih.gov/Programs/Pages/biodiversity.aspx, accessed 27/9/2013.

International Cooperative Biodiversity Groups (ICBG). '*ICBG Madagascar*'. Available at: http://www.icbg.org/groups/madagascar.php, accessed 21/8/13.

Kingston, D.G.I. (2011) 'Modern natural products drug discovery and its relevance to bio-diversity conservation'. *Journal of Natural Products,* 74(3), 496–511.

MBG Madagascar Research and Conservation Program. (2012) *Annual Report 2012.* Anta-nanarivo, Madagascar.

Neimark, B.D. (2012) 'Industrializing nature, knowledge and labour: The political econ-omy of bioprospecting in Madagascar'. *Geoforum,* 43(5), 980–990.

Rakotonjatovo, B.H., Rasolofoarimanga, H.A., Andriamanantoanina, H., Ranarivelo, L., Maharavo, J., Ramaroson, L. and Ratsimbason, M. (2006) 'A microfluorimetric method to screen marine products for antimalarial activity – preliminary results'. In Midiwo, J.O., Yenesew, A. and Derese, S. (eds) *Proceedings of the 11th NAPRECA Sympo-sium.* Antananarivo, Madagascar, pp. 154–160. Available at: http://www.napreca.net/ publications/11symposium/pdf/Q-154–160-Ratsimbason.pdf, accessed 17/10/2013.

Ratsimbason, M., Ranarivelo, L., Juliani, H.R. and Simon, J.E. (2009) 'Antiplasmodial activ-ity of twenty essential oils from Madagascan aromatic plants'. In Juliani, H.R., Simon, J.E. and Ho, C.T. (eds) *African Natural Plant Products: New Discoveries and Challenges in*

Chemistry and Quality Control. American Chemical Society Symposium Series 1021. American Chemical Society, Washington, DC, pp. 209–215.

Rosenthal, J. (1997) 'Equitable sharing of biodiversity benefits: Agreements on genetic resources'. In *Investing in Biological Diversity – Proceedings of the International Conference on Biodiversity Incentive Measures.* OECD Press, Paris, pp. 26.

William L. Brown Center of the Missouri Botanical Garden (WLBC/MBG). '*Madagascar*'. Available at: www.wlbcenter.org/madagascar.htm, accessed 17/10/2013.

4   ABS agreements in Thailand

This chapter focuses on 2 existing ABS agreements and the national legal context in which they have been developed. Although implementing rules under one of the national laws had not been passed until very recently (in mid-2013), agreements had been established with the permission of the relevant department (in the Shiseido case) and through a contract (in the Novartis case) according to mutually agreed terms (MAT) (in both cases). Monetary benefits are yet to be shared as research is continuing on the cosmetic and medicinal potential of the respective genetic resources. Having said that, it is worth examining the agreements as there are already some non-monetary benefits to be discussed from the collaborations that have been established.

There are 2 agreements of interest here, both with the National Center for Genetic Engineering and Biotechnology (BIOTEC) of Thailand. BIOTEC is one of 4 technology centres under the auspices of the National Science and Technology Development Agency (NSTDA), which is an autonomous government agency. The centre has over 30 laboratories and 150 principle scientists conducting basic and applied research, heavily focused on microorganisms, on a wide range of applications such as agricultural, biomedical and environmental sciences. The Center also acts as a repository for microbes, providing a cell culture storage facility as an ex situ storage service for researchers, companies and for the Thai government. This ex situ facility is itself an important conservation measure. The microbial culture collection has over 50,000 strains and the fungal herbarium has over 30,000 samples, which have been collected from throughout Thailand in the past 20 years. BIOTEC also has a chemical library of pure compounds and chemical profile databases of crude extracts, mainly from fungi and actinomycetes (BIOTEC, 2013).

The partners claim that the agreements have occurred in compliance with the CBD and, in the Shiseido case, Thailand's Plant Varieties Protection Act B.E.2542 (1999) (hereafter, Thai PVP Act). To provide some background, the Thai government signed the CBD in 1994, but did not ratify until 2003. During the interim between signature and ratification, considerable inter-agency negotiations and consultations occurred regarding the most appropriate regulatory framework for ABS. At the same time, there were a number of media reports and some NGO activism over a number of cases perceived to be misappropriations or 'biopiracy' of Thai genetic resources and traditional knowledge (TK) (see Robinson, 2010 for detailed accounts). Given that Thailand has many unique varieties of agricultural crops (e.g. varieties of Jasmine rice) that have received considerable improvement through farmer and institutional plant breeding, and that make a major contribution to agricultural exports, a specific agriculture-focused law was developed: the Thai PVP Act (administered by the Department of Agriculture (DoA)). Simultaneously, an Act on the Protection and Promotion of Thai Traditional Medicinal Intelligence was developed by the Department of Public Health to protect and promote the considerable array of herbal and traditional remedies, tonics and treatments in the country. Since the year 2000, a number of regulations and rules under these 2 laws have been considered by the Council of State, with only some being passed to date (see Robinson and Kuanpoth, 2009). Only very recently (mid-2013) have the rules dealing with access and benefit-sharing, which ultimately have fallen under the jurisdiction of the DoA, been approved. In the meantime, the DoA has been able to provide permission to collaborative biodiscovery activities according to 'letters of intent' to establish benefit-sharing upon commercialisation.

The 2 agreements, with the Japanese cosmetics company Shiseido, and with Swiss multinational healthcare company Novartis, are each discussed in turn.

The BIOTEC – Shiseido collaboration

Access and R&D conducted

The joint-collaboration between BIOTEC and Shiseido started in December 2004. Information including a plant list and research plan was submitted to the DoA and permissions were granted (Kirtikara, cited by BIOTEC, 2008). The research had long-term aims of analysing the potential use of plant extracts as cosmetic ingredients, as well as research towards product development and commercialisation. It involved a professor from Mahidol university, who had established a database of plants with medicinal use, with associated scientific publications (http://www.medplant.mahidol.ac.th). This was searched using a number of specific keywords of interest to Shiseido to detect potential dermatological and hair applications of the plants.

Following this scoping phase and literature reviews of plant information, 103 plants were searched and then 15 were selected to perform in vitro screening and tests for safety, efficacy and formulation. In the initial stages of the research, local knowledge of plants was not necessarily used as a lead in the discovery process, but rather evidence from public domain publications was used. Only during later discussions regarding the shortlisted plant candidates of interest did stories about past 'traditional' uses (as medicines, not cosmetics) come up (BIOTEC and Shiseido staff, interview, 19/6/2013).

Plant extracts (usually dried raw leaves, bark or other plant matter) were obtained from various sources, including local apothecaries which are common throughout Thailand. These were transferred to Japan under the partnership agreement which covered not only the matter of material transfer but also other elements such as the scope of work and responsibility of each party. In Japan, the majority of the screening was undertaken, according to specific terms and restrictions outlined in the agreement (e.g. restrictions on third-party transfer). The first memorandum of understanding between BIOTEC and Shiseido was relatively broad in terms and scope, focusing on the intent of the partners to form a research collaboration, and thus a second agreement was established prior to the filing of patents, with the partners reaching MAT regarding mutual research benefits, and making commitments to make new agreements on a case-by-case basis when the potential for publications or joint-patents arises. The second agreement established the intent to conduct an in-depth study of plant extracts that can potentially be used as cosmetic ingredients. Terms and conditions on transfer of biological materials were incorporated into the second agreement, as one of the articles. This Agreement was signed before the research project commenced. A third agreement was then made towards a joint patent application: this agreement sets out the terms and conditions on the joint patent on in-vitro activities of 6 plant species, in Japan and Thailand (BIOTEC and Shiseido staff, interview, 19/6/2013).

The transfer of plant materials to Japan was processed under all applicable laws in Thailand (i.e. the Thai PVPA and Plant Quarantine Act B.E. 2507). At that time the Ministerial Regulation Section 53 on ABS of the Thai PVPA was already enforced, but Section 52 was still under development. This collaboration started with research intention, therefore, Section 53 applied. When both parties agreed to seek for patent protection on their work, BIO-TEC asked for advice from the DoA. The PVP officer reported that filing for IP protection has been regarded as commercial purpose. Therefore Section 52 needed to be complied with. However, the Ministerial Regulation was not in place yet, and so the agreement between DOA and BIOTEC came in the form of 'letter of intention' (BIOTEC and Shiseido staff, interview, 19/6/2013).

A commitment was made by BIOTEC and Shiseido through the 'letter of intent' to the DoA to establish a benefit-sharing agreement once the implementing rules under the PVP Act were put into place. Now that the rules have been passed in mid-2013, the parties expect that they will have to meet again soon to discuss a benefit-sharing agreement.

By October 2006, Shiseido had opened its Southeast Asian Research Center, located in Bangkok close to BIOTEC headquarters to foster joint-projects between the partners (BIOTEC, 2008). The partners stressed that this initial project was one primary factor which had led to the establishment of this regional research center – seen as an important investment and technology transfer outcome for Thailand (BIOTEC and Shiseido staff, interview, 19/6/2013).

By July 2008, their joint research had identified potential cosmetic applications through in vitro testing of 6 plant species including:

- *Cinnamomum ilicioides* (possessing whitening effects, anti-aging and skin smoothing activity);
- *Lepisanthes fruticosa* (possessing anti-aging and slimming effects);
- *Fagraea fragrans* (possessing anti-aging and skin smoothing effects);
- *Pterocarpus indicus* (possessing anti-aging effect);
- *Boesenbergia regalis* (possessing whitening and anti-aging effects); and
- *Aquilaria crassna* (possessing whitening and anti-aging effects).

As a result, the collaborators filed for joint-patents in both Thailand and Japan. While these are difficult to identify for Anglophones through these 2 respective national patent offices (they are filed in Thai and Japanese), an additional filing has been made to the World Intellectual Property Organization (WIPO) under the Patent Cooperation Treaty (PCT). The patent, as it appears in WIPO, has a Publication Number WO/2009/157575 (Publication Date 30/12/2009) and an International Application Number PCT/JP2009/061796 (International Filing date 22/6/2009), and title 'Skin preparation for external use, whitening agent, anti-aging agent, and antioxidant'. The patent appears as a composition or mixture made up of 'several kinds of plant extractions' made up from the above list of plant species. Shiseido and the NSTDA are the applicants for all designated states.[1] The designated states include all major patent offices including the European Patent Office (EPO), the United States and Canada (and it was filed via the Japanese Patent Office). This international patent application has been filed to obtain a priority filing date (i.e. to be first to file an application of this kind).

Benefit-sharing

The parties to this agreement stressed that they had always set out to establish their research together as a collaborative partnership in compliance with the ABS principles set out under the CBD. The primary benefit established through this collaboration has been technology transfer and the sharing of research results, data, training and seminars (non-monetary benefits recognised in the Annex to the Nagoya Protocol). BIOTEC noted that they had benefited from new types of screening assays under the partnership, for example, and that there had been numerous laboratory visitations between the partners. The establishment of the Shiseido Southeast Asia Research Center as a subsequent result of this collaborative partnership can be considered an important additional benefit, and arguably this investment has both monetary and non-monetary benefits for Thailand. Subsequently, BIOTEC and Shiseido have started an additional research project which involves the analysis of microbial diversity on facial skin.

The jointly owned patent applications are another obvious example of benefit-sharing to date. They are listed in the Annex of the Nagoya Protocol and considered as both monetary and non-monetary benefits. Upon commercial exploitation of the protected product, the parties that hold a joint-patent often agree to a division of royalties from licence fees or similar. The joint-ownership also allows for non-monetary benefits such as the recognition that a patent provides to researchers and institutions, and the ability to jointly make decisions over the exploitation of those patent rights.

Impact

Prof. Chachanat Thebtaranonth, NSTDA Vice President and Executive Director of the NSTDA Technology Management Centre, has indicated that 'It has long been NSTDA's interest to collaborate with foreign partners in research and development. Our collaboration is based on the partnership concept, in which both partners co-invest and share results and benefits. NSTDA also gives importance to enhancing national research capabilities' (BIOTEC, 2008). Although there are not yet commercial products or monetary benefits, there is still reasonable potential for these to arise. Given that the Shiseido research centre has only been in place since 2006, there is potential for other collaborations with BIOTEC or other Thai institutions in the future, discussed below.

Discussion

There may be implications and spin-offs as a result of the collaboration. BIO-TEC has been cited as saying that it hoped the joint project would encourage the use of Thai herbs in cosmetic products across the globe. Pitman (2005) noted that Thailand is already one of the world's leading suppliers of herbs to cosmetic companies, with some 10,000 varieties of herbs grown in the country, giving the industry an estimated value of THB 30 billion (approximately $985 million). If successful, the joint-venture may lead to further collaborative research efforts and may stimulate additional commercial interest from third parties.

The patent of interest here, as it appears in the WIPO database with title 'Skin preparation for external use, whitening agent, anti-aging agent, and anti-oxidant', has commonalities with many other cosmetic patents. The purpose is the same (various improvements to skin appearance), it is based on a mixture composed of several plant extracts, and it seeks to achieve its stated purpose through a novel mixture. While some of the press releases from BIOTEC highlight that the intent of the joint research was to identify new cosmetic uses from existing or known Thai plant varieties (many of them used for other medicinal or food purposes), the inventiveness of the patent application has been raised in one assessment. A 'Written Opinion of the International Searching Authority' report under the PCT notes there are problems with the 'inventive step' claims of the patent for all 5 claims made. They cite several Japanese and Korean patent documents that indicate similar activity from the same or similar plant extracts – this is a common problem with patent applications for mixture-type cosmetic applications. Unless synergistic activity can be proved, and this can be quite difficult, it can be hard for the patent examiner to accept that the claims are indeed 'inventive'. Without synergistic bioactivity, it can appear just like a conveniently new mixture of different known ingredients. Ultimately, the grant of patents will be determined by national patent offices. BIOTEC noted that it was important to file the patent under the PCT so as to establish a priority filing date, thus staking the claim in 2009 to this invention.

At least 2 of the plant species are quite commonly known to be used as traditional medicines in Thailand. Although a cosmetic application does appear to be new in each case, sometimes the medicinal applications can provide a lead towards cosmetic bioactive ingredients. For example, *Fagraea fragrans* (Roxb.) is one of the species described in the patent and can be found in Saralamp et al. (1996) *Medicinal Plants in Thailand*. The description from this text is that the leaf has uses as an 'antimalarial, element tonic, antiasthmatic; externally used for mild infectious skin diseases'. So there is clearly some existing prior knowledge of this plant for medical skin treatments in Thailand, although probably not explored to determine its safe use, effective quantities or similar information relevant to its development.

Importantly, BIOTEC/NSTDA and Shiseido have made commitments regarding the impact of the pending patent. The partners have made it explicit in the patent documents and in press releases that Thai local communities can continue using the plant extracts for a variety of purposes, as they have traditionally been used. The Executive Director of BIOTEC has said, 'This is an unprecedented move in Thailand for a patent to give favour to local communit[ies], and we hope it will set [an] example for future practices, not only to protect our biodiversity, but also to give recognition to our indigenous knowledge' (BIOTEC, 2008).

While these commitments are important, it must be recognised they are also really assurances of an existing natural right for Thai local communities to use plant extracts for a range of purposes.[2] Patent rights are limited in scope to the claims being made – in this case what appears to be a new mixture for cosmetic use, and this scope is only enforceable in the countries in which a patent is granted. At this stage no patent has been granted. If the patent were granted in Thailand, one concern might have been regarding existing herbalists and small companies that use these plant-based ingredients for skin treatments (such as that described above in Saralamp et al., 1996). There have been cases in the past, including some in Thailand, where the attorneys representing a patent holder have sought to over-inflate the scope of a patent and to intimidate local businesses who rightfully had been dispensing herbal medicines for decades (see Robinson and Kuanpoth, 2009). Some of these instances may have been motivating factors for the development of the 2 laws mentioned earlier. Representatives of both BIOTEC and Shiseido were acutely aware of these sorts of past controversies and had always set out to approach their partnership in accordance with the ABS principles outlined in the CBD. Given the goodwill already established in this partnership, and the provisos put in place to ensure local use is allowed in the patent documents, the prospects of any such problems occurring in this case seem extremely low.

The BIOTEC – Novartis collaboration

Access and R&D conducted

In March 2005 Novartis signed a 3-year partnership agreement with BIO-TEC for the analysis of natural products including micro-organisms and natural compounds for the development of new medicines. The research is targeting diseases such as cancer, heart disease, diabetes and tropical diseases, according to Dr Daniel Vasella, Chairman and CEO of Novartis (Jager, 2005). This is one of only a handful of remaining natural products ventures by Novartis, with much of their research focus shifting to combinatorial chemistry, recombinant biotechnology and synthetic biology (among others) in recent years.

The partners point to the global and conservation benefits of the research, noting that only 1.9 million species of living organisms are recorded by scientists, compared with conservative estimates of about 12 million species waiting to be discovered (up to about 30 million). Specifically on fungi, Dr Morakot, the former Director of BIOTEC estimates that approximately 100,000–150,000 fungi would be hosted by Thailand, roughly 7–10 per cent of the global estimation. But barely 1 per cent of this presumed total has actually been described. One of the common spin-off benefits of natural products drug discovery research is the identification and characterisation of new species. This can have important implications for understanding the contribution of these species to ecology and ecosystem function (e.g. carbon and nutrient cycling).

The collaborative arrangement has been managed by a Joint Steering Committee consisting of scientists and executives from both parties. This initial 3-year partnership was extended in 2008 for a further 3 years from June 2008 to May 2011, and then again for 3 years from 2011. By the end of the first 6 years of the partnership, 7,200 microbial isolates and 115 pure compounds from Thailand have been investigated through a series of drug tests covering disease areas under the Novartis research portfolio, such as infectious and cardiovascular diseases, oncology and immunology (Murarka, 2012).

There is no law in Thailand regulating the utilisation of microbes for R&D. The Act on Protection and Promotion of Thai Traditional Medicinal Intelligence does not apply in this case either. Instead, the transfer of microbes and compounds was made under a material transfer agreement with specific terms regarding use, third-party transfer, intellectual property and benefit-sharing. The microbial strains remain the property of BIOTEC, and Novartis receives a time-limited, exclusive user right. Novartis are able to track which compounds have been isolated from which genetic resource through a natural products database (Peterson, 2013). BIOTEC is also allowed to use the same strains for their own research simultaneously.

The microbes that were transferred to Novartis were only those that were specially collected by a BIOTEC team for the collaboration, and not specific collections transferred to and held by BIOTEC for other institutions and researchers. This special collection has not been made public because of the exclusive 'Cooperation and Commercialization Agreement' between BIOTEC and Novartis, but they may be upon the end date of the agreement. When collected from specific areas,

BIOTEC has procedures in place to ensure that permissions are obtained for the activity. For example, they often collect samples in National Parks and are required to obtain permissions for research prior to entry and sampling under a law administered by the Thai Department of National Parks, Wildlife and Plant Protection. When sampling in marine areas there are only sometimes permit requirements for marine protected areas (BIOTEC staff, interview, 19/6/2013).

Benefit-sharing

One major component of benefit-sharing is through an internship arrangement whereby Thai researchers are able to work at Novartis' laboratories in Switzerland, as well as having Novartis' scientists visit BIOTEC and share their knowledge. This has led to improvements of research capacity in BIOTEC including:

- The chemistry lab is able to automate and improve extraction of compounds and chemical screening processes, with equipment such as High Performance Liquid Chromatography – Mass Spectrometry subsequently purchased by BIOTEC to improve the automated screening process;
- Two new bioassay screenings have been added (an anti-tick assay and a sensitised *Staphloccoccus aureus*-based antibiotic assay) and can now be offered as part of BIOTEC's range of screening services (Murarka, 2012);
- The establishment of new directions for BIOTEC's research programs, such as the identification of new types of micro-organisms (actinomycetes and myxobacteria). In the first few years of study over 2,000 actinomycetes from 25 genera were isolated from Thailand. Actinomycetes are considered to be a very important source of natural antibiotics.

In addition to these non-monetary benefits, some BIOTEC scientists have been invited to talks in Geneva to present on this collaboration. BIOTEC has also received Novartis scientists in Bangkok, who have given seminars from time to time.

While there have been no discoveries warranting joint patents, the program is still relatively new considering how long the drug-related R&D process can be. Because of the commercial nature of the research there are no joint-publications – if these were to occur, they would likely be if or when patent approval was obtained.

In the agreement with Novartis, there are specifications regarding milestone payments at various drug development and clinical trial stages. To date no potential drug candidates have been identified through the screening process and so no milestone payments have been made. These amounts were not specified and remain confidential.

Impact

To date there have been 9 scientists from BIOTEC who have taken up the opportunity for a 3-month internship in Basel, at Novartis' laboratories. This training and technology transfer has been an important opportunity according to BIOTEC staff. Up to 3 scientists per year are eligible to undertake the internship under the agreement, however not every year are this many staff able to take time out from their normal work and lives for the internship.

A Thai PhD student will soon graduate in the field of actinomycetes, under the joint supervision of a Thai professor and a Novartis scientist (now retired from Novartis, but still supervising). These sorts of collaborations have led to new research activities and projects that BIOTEC might not have been able to undertake otherwise. The staff at BIOTEC highlighted that this had been important in terms of public recognition of their research (which is also considered a non-monetary benefit in the Nagoya Protocol Annex).

Discussion

Given that the research is focused on micro-organisms and natural compounds for medicinal or pharmaceutical purposes, there are always long periods of R&D prior to commercialisation of a new product. This means that it is difficult to expect any significant monetary benefits up-front or even several years after the joint-project has been running. However, collaborative research arrangements under the ABS framework established by the CBD (and as guided by the Bonn Guidelines and Nagoya Protocol) can include the exchange of non-monetary benefits and training. Indeed this agreement has been described as a potential model of biodiversity research collaboration.

In addition, BIOTEC staff noted that it is important to highlight that ABS is not always a 'south-provider, north-user' transfer of genetic resources and knowledge. Indeed this sort of partnership has meant that a 2-way exchange of R&D and knowledge has occurred, alongside the transfer of microbes from biologically diverse Thailand to Switzerland. The establishment of a Regional Biotechnology Training Center by BIOTEC has also led to further south-south training of scientists from Southeast Asia and from the Pacific. This has inevitably benefited from BIOTEC's partnerships, and since the inception of the partnership, Novartis has stated that it wishes to assist Thailand to become a 'center of excellence' for natural products research and discovery in the region.

Another Novartis example in brief

As noted earlier, there are few large pharmaceutical companies still actively engaged in bioprospecting activities or natural products drug discovery. One of the only large companies that still retain a natural products research division is Novartis. Apart from the partnership that Novartis has with the Thai BIOTEC authority, they also have another 2 partnerships of interest: with the Shanghai Institute of Materia Medica (SIMM), since 2001, and with the Sarawak Biodiversity Center (SBC), since 2009.

The Novartis Headquarters in Basel were contacted in relation to these (10/7/2013), particularly the more well-established partnership with SIMM.

Frank Petersen, Executive Director of Natural Products, responded with a short response to several questions. He described Novartis' natural products research program as focusing on:

- Accessing new micro-organisms as sources for new natural products, and access to purified plant metabolites through partnership programmes;
- Identification and isolation of biologically active natural products from plants and micro-organisms;
- Identification of new natural products for biological screening;
- Genetic engineering for identification of new natural products and yield optimisation;
- Investigation of secondary metabolism;
- Fermentation technologies;
- Supply of larger quantities to support [medicinal chemistry], pharmacology, toxicology, formulation, and stability;
- Elaboration of manuals for "Good Manufacturing Process" to enable clinical research (pers. comm. 17/9/2013).

Regarding the relationship with SIMM, Novartis was contacted by their Chinese partners and then analysed the scientific and technical expertise at SIMM. This led to a contract signed by SIMM as a 'legally entitled institute'. SIMM received technology transfer and training in the newest analytical and preparative technologies, 8 visiting scientists were trained at Novartis in Basel (including full cost coverage) (Petersen, pers. comm. 17/9/2013). Several thousand compounds were received from SIMM and screened in HTS projects at Novartis, and there have been scientific publications as a result but no follow-up R&D (pers. comm. 17/9/2013).

Petersen explained that there was a benefit-sharing agreement in place with prior informed consent and MAT. However, he did not elaborate further on these terms, nor does Novartis' website or related publications explain any further. He also noted that there is no bioprospection of plants conducted, natural products produced by endangered plant species are excluded, and that substances are selected based on the 'novelty of structures' (pers. comm. 17/9/2013), suggesting that TK is not used as a lead for their R&D.

Notes

1 Except for the United States which requires named persons, in which case the patent also lists several authors, assumedly from Shiseido Japan, and 2 authors from Thailand including the Executive Director of BIOTEC.
2 There are some restrictions on plant collection under regulations to the Act on Protection and Promotion of Thai Traditional Medicinal Intelligence – a list of threatened and rare species of medicinal plants can been scheduled by the Department of Public Health. To date only one plant species has been listed under the regulations – *Pueraria mirifica*.

References

BIOTEC. (2008) '*Joint Patent for Research Collaboration between BIOTEC/NSTDA and Shiseido. Press Release*'. Available at: www.biotec.or.th, accessed 13/12/2012.

BIOTEC. (2013) '*Research Overview*'. Available at: http://www.biotec.or.th/en/index.php/research/overview, accessed 11/7/2014.

Jager, P. (2005) '*Novartis and BIOTEC Sign Three-Year Research Collaboration. Media Release*'. Available at: www.novartis.co.th, accessed 13/12/2012.

Murarka, V. (2012) 'Novartis and biotec tie-up to explore Thailand's biodiversity'. *BioSpectrum Asia.* Available at: www.biospectrumasia.com/biospectrum/influencers/1279/novartis-biotec-tie-explore-thailands-biodiversity, accessed 30/4/2014.

Petersen, F. (2013) 'Accessing microorganisms as genetic resources for natural products in drug discovery'. Presentation 5 February 2013, IFPMA Side event, WIPO IGC meeting 23.

Pitman, S. (2005) 'Shiseido eyes Thai Herbal ingredients'. *Cosmetics Design Asia,* Available at: www.cosmeticsdesign-asia.com, accessed 29/5/2012.

Robinson, D. (2010) *Confronting Biopiracy: Cases, Challenges and International Debates.* Routledge/Earthscan, London.

Robinson, D. and Kuanpoth, J. (2009) 'The traditional medicines predicament: A case study of Thailand'. *Journal of World Intellectual Property,* 11(5/6), 375–403.

Saralamp, P., Chuakul, W., Temsiririrkkul, R., Clayton, T. and Paonil, W. (Vol 1 – 1996; Vol 2 – 1997) *Medicinal Plants in Thailand.* Mahidol University, Bangkok.

5   The Samoan Mamala case

Although this chapter refers to events that began prior to the establishment of the CBD, there are relevant actions undertaken by the researcher and subsequent research institutions of relevance to ABS. This case study examines the discovery of an anti-viral phorbol (prostratin) from ethnobotanical study of Samoan remedies in the late 1980s, and the agreements put in place to benefit the community of Falealupo and the people of Samoa. Prostratin was identified by Dr Paul Cox of the Institute for Ethnobotany as an isolated extract from traditional healer remedies that used the rainforest tree known locally as 'Mamala' (*Omolanthus/Homalanthus nutans*). Three agreements are relevant:

- The Falealupo Covenant which allowed Dr Paul Cox to access a community-held rainforest area for biodiscovery purposes (1989).
- An AIDS Research Alliance (ARA) agreement with the Government of Samoa (2001).
- A University of California, Berkeley agreement with the Government of Samoa (2004).

The access, biodiscovery and research activities

Dr Paul Cox conducted ethnobotanical studies in the Falealupo rainforest, reputedly collecting many plants. Mamala was identified as a plant of interest during discussions between Dr Cox and some of the village healers (see Figure 5.1). According to Dr Cox's field notes:

> Several healers, including Epenesa Mauigoa, Pela Lilo, and Seumantufa's wife Lemau, told me that water infusions of *Homalanthus* are used to treat yellow fever[1] and intestinal complaints . . .
>
> (Cox, 2001, p. 35)

Cox found that although there was broad knowledge in Samoa of the use of *Homalanthus* to treat intestinal complaints (the use indicated by Lemau Seumantufa), only 2 healers, for example, Epenesa Mauigoa and Pela Lilo, knew of the use of the plant to treat acute viral infection (pers. comm. 6/6/12).

Interviews with a number of members of the community confirmed Dr Cox's ethnobotanical activities in the late 1980s, including an interview with Lemau (noted above, interview, 23/5/12). However a number of people now claim that the knowledge relating to Mamala might be held by many traditional healers in Samoa and this is described by Whistler (1996, 2004).

Subsequently, Dr Cox received an invitation from the US National Cancer Institute (NCI) in 1986 to collaborate in screening the plants for anti-cancer activity. While they did not identify activity against cancer, subsequent testing by the NCI identified the prostratin molecule displayed 'potent cytoprotective activity' – the capacity to protect healthy cells from a range of pathogens. In vivo studies with mice also showed that, despite being a phorbol (which often promote tumour growth), prostratin did not promote tumours (Cox, 2001). Further research has been ongoing on the use of prostratin as a potential anti-retroviral pharmaceutical to treat HIV. As the ARA notes:

> Prostratin is a member of a distinct subclass of protein kinase C (PKC) activators. Activation of the PKC pathway triggers a cascade of events that result in the transport of important transcription factors that ultimately "turn on" viral gene expression. A variety of studies conducted by AIDS Research Alliance and other scientists have confirmed the specific activity of prostratin and its effects within the cell. Several reports have demonstrated prostratin's potent antiviral activity against various strains of HIV.
>
> (ARA, accessed, 21/1/14)

Figure 5.1 The 'Mamala' tree at Falealupo, Samoa

Source: Robinson, 23/5/12.

ARA has since been conducting pre-clinical studies on prostratin with funding from the US National Institutes of Health (NIH). Research has not yet entered phase 1 clinical trials.

Access/prior informed consent procedure followed

Dr Cox (2001) explained that having previously spent considerable time living in Samoa (and having learnt the language), he returned with funding from the US National Science Foundation to conduct an ethnobotanical study in the mid-1980s and settled in the village of Falealupo on Savaii. Dr Cox indicated that he initially obtained verbal prior informed consent for his ethnobotanical studies:

> My first introduction to the village was a kava ceremony with the village chiefs where I explained the purpose of my research, and asked their per-mission to study with the village healers and to collect their medicinal plants for laboratory analysis. I also told them that there was a slight chance that a discovery could result in a commercial interest, and pledged to do my best to ensure a return to the village from any discovery. The village chiefs unanimously agreed to grant me permission to conduct the research and to assist me in any way that they could.
>
> (Cox, 2001, p. 35)

This research was prior to the CBD, and therefore ABS principles had not yet been developed. There does not appear to have been a specific research permit requirement in Samoa at the time – many countries have implemented such requirements following ratification or accession to the CBD. Most of my interviews in Falealupo concurred with this depiction of events; however, one interviewee indicated that Dr Cox had taken some medicines when he first came to stay, before the Falealupo Covenant.

Following the verbal admission of consent from the Falealupo chiefs, Dr Cox continued his research in the Falealupo forest from 1985 to 1988. In 1988 the Samoan Government required the Falealupo community to build a primary school for the community, which had led to a deal with a logging company. The company began felling trees in the local forest to raise sufficient income for the school construction. In late 1988/early 1989, Dr Paul Cox (2001) met with the community to negotiate a covenant for the protection of the forest and repayment of the community's $85,000 debt, in exchange for continued ethnobotanical research access. Several interviews confirmed this agreement was made, that the community was happy with the payment of the debts and establishment of the rainforest reserve (Fuiono Patolo, interview, 23/5/12; Seumantufa Fale mai, interview, 23/5/12; Manu Toifotino, inter-view, 24/5/2012; Taii Tulai, interview, 24/5/2012) (Figure 5.2 is a photo of the rainforest preserve). Dr Cox was subsequently titled 'Nafanua', which is an honorary title named after a goddess warrior from Falealupo who conquered and united Samoa.

Figure 5.2 The Falealupo rainforest from the top of the Canopy Walk

Source: Robinson, 15/3/2012.

The Falealupo Covenant described briefly that the above was signed by every chief in Falealupo in a kava ceremony attended by the village (with the exception of one 'banished' family, as noted by several current chiefs).[2] One interviewee noted that the available text of the covenant had changed somewhat with re-draftings and translation, changing some of the meanings afterwards (interview, 23/5/12). Dr Cox (2001) also met with the Samoan Prime Minister Tofilau Eti, the Samoan Minister of Agriculture Solia Papu Va'ai (confirmed by interview in March 2012) and a number of members of parliament to notify them of the preliminary research findings and the NCI's commitment to both honour the Falealupo Covenant and to require any licensee to negotiate fair and equitable terms of benefit-sharing of any proceeds arising from a patent (issued later in 1996). Dr Cox later was involved in the negotiation of 2 memoranda of understanding establishing terms for potential benefit-sharing between the Samoan Government, the ARA and University of California Berkeley. These negotiations involved Mr Solia Papu Va'ai, who was the Member of Parliament for Falealupo at the time. It is unclear to what extent the Falealupo community was involved in negotiation of these 2 agreements. According to Cox, he took representatives from both the ARA and the University of California to Falealupo where the village chiefs reviewed and signed the respective benefit-sharing agreements (Cox, pers. comm. 6/6/2012). My interviews in Falealupo with several chiefs could not identify their involvement in the negotiations.

Terms of benefit–sharing agreements

Falealupo Covenant: the main terms in summary include:

- The payment of a debt of $77,000 WST to the Bank and $31,000 WST to Samoa Forest Products for the construction of the school (total of approximately $85,000 at the time),
- An acknowledgement of the perpetual sovereignty of the Falealupo community over the rainforest,
- A commitment by the community to preserve the rainforest for 50 years, including limitations on hunting and allowance for traditional uses,[3]
- Allowance for Dr Paul Cox and his associates to access the rainforest for scientific research in perpetuity, as long as they do not damage the rainforest. If Dr Cox is successful in finding new drugs, he will return to the village 33 per cent of the income received.

It is worth also noting that Dr Cox and his associates subsequently established the NGO Seacology which has provided a number of benefits to the Falealupo community including:

- The construction and maintenance (costing over $100,000) of a rainforest walkway which generates ecotourism income for the community,
- The re-construction of the school, water tanks and a clinic, plus emergency supplies after hurricanes Ofa and Val (costing over $160,000),
- Plus subsequent personal contributions by Dr Cox to a perpetual endowment fund and village retirement funds.

Dr Cox (2001) estimated that over $480,000 in contributions have been made to Falealupo village by himself, his associates and Seacology. Interviews with some members of the community in May 2012 confirmed that the approximate individual costs of specific contributions would tally to this approximate amount, though some noted that an 'environment centre' which has been mentioned as provided, never eventuated in the village (Seaumantufa Falemai, interview, 23/5/12; Manu Toifotino, interview, 23/5/12). Some interviewees who were critical of the agreements suggested that Cox would not have received the recognition and prizes that he had if it wasn't for the community. On the other hand, Cox donated his share of the 1997 Goldman Environmental Prize to Seacology to create an endowment for benefit of the Falealupo Forest. These funds have been used to pay for annual maintenance on the Falealupo Rainforest Canopy Walkway, and, prior to the stock market collapse in 2008, annual payments to the village (Cox, pers. comm. 6/6/2012). From the interviews conducted there was little knowledge about the way the perpetual fund operates, and a number of community members noted that annual payments had recently stopped (Taii Tulei, interview, 23/5/12; Fuiono Patolo, interview, 23/5/12).

The ARA – Government of Samoa Agreement

Subject to a number of qualifications such as the passing of 3 phases of testing and US FDA approval, and the capacity for ARA (a not for profit) to partner with a company and generate surplus revenue net of expenses for ARA, the ARA agreed to pay the following to the Samoan Government:

- $5,000 as a good faith payment,
- $10,000 as a milestone payment for the passing of phase 1 trials,
- $20,000 for passing phase 2 trials,
- $40,000 for passing of phase 3 clinical trials, and
- Once passed USDA approval if ARA realises revenues that exceed costs, they agree to pay royalties of this revenue at:
    - 12.5 per cent to the Samoan Government;
    - 6.7 per cent to Falealupo village;
    - 0.4 per cent to the lineal descendants of Epenesa Mauigoa, late of Pesaga village (being the first healer to identify for Dr Paul Cox 'mamala' as having potential anti-viral qualities); and
    - 0.4 per cent to the lineal descendants of Pela Lilo, late of Falealupo village (who shared ethnobotanical knowledge concerning formulation and use of mamala as a treatment for illness).

While interviewed members of the community had some knowledge of this agreement, noting that Dr Cox had visited and discussed it in 2001, they only had basic knowledge of the division of potential royalties (Taii Tulei, 23/5/12; Fuiono Patolo, 23/5/12; Manu Toifotino, 23/5/12). Seaumantufa Falemai and his mother Lemau were surprised that they were not listed as direct beneficiaries in this agreement (interviews, 23/5/12 and 24/5/12). Lemau indicated that she was part of the first family where Cox stayed, and that she was told she would benefit directly from the ARA agreement (interview, 23/5/12). However, Cox indicates that although Lemau taught him many things about Samoan healing plants, she did not teach him about the use of *Homalanthus* to treat viral illness and only used it to treat intestinal complaints (Cox, pers. comm. 6/6/12).

Some interviewees and Samoan attendees at workshops in the Pacific essentially noted that this was a 'pointless agreement' because there would never be any profit for the 'not-for-profit' ARA. They also questioned the 'fairness and equity' in the (potential) distribution of benefits (interviews and meeting notes, 22–25/5/2012). One interviewee noted the existence of a Samoan Society of Traditional Healers – there was the suggestion that this could be one avenue for benefit-sharing.

The UC Berkeley – Government of Samoa Agreement

Subject to a number of terms, 50 per cent of the net revenue proceeds (after reimbursing costs and legal fees) that arise from UC Berkeley's licensing of intellectual property directly from this research on *Homolanthus nutans* will go to the NGO Seacology to be distributed as follows:

- 50 per cent to the Samoan Government,
- 33 per cent to Falealupo village,
- 2 per cent to Saipipi village,
- 2 per cent to Tafua village,
- 8 per cent to other villages that participate in the research or allow collection of mamala, grow crops of mamala at the time of FDA approval of prostratin or its analogues as a drug in a reasonable and equitable manner as decided by Seacology,
- 2 per cent to the lineal descendants of Epenesa Mauigoa,
- 2 per cent to the lineal descendants of Pela Lilo, and
- 1 per cent to Seacology for handling the royalty payments.

Several interviewed members of the community had only limited awareness of this agreement, and very little knowledge of the terms or progress of the R&D. Again the likelihood and equity of benefit flows was questioned by some Samoan delegates in our ABS meetings. Related to this, and given the large proportion of proceeds to the Samoan Government, some letters in Samoan newspapers had been quite critical of the Government's 'selling off of Samoan resources' and of potentially missing out on licensing fees from use of the patents (see the *Samoa Observer*, Palamo, accessed 16/5/2012).

Summary of monetary and non-monetary benefits

The community has received considerable monetary benefits from the initial Falealupo Covenant, the subsequent philanthropic contributions made by Dr Cox and Seacology, and there are considerable terms of royalties and milestone payments established in the subsequent agreements (although almost entirely contingent upon successful commercialisation) (Table 5.1).

Tangible impacts derived from these benefits

Although royalty payments 'arising from the utilisation of genetic resources' have not yet been made, a number of benefits for Falealupo community are visible. These have been made under the Falealupo Covenant (much like an 'upfront/access fee') and through philanthropic donations made by Dr Cox and Seacology.

From several interviews, every respondent indicated that they were happy that the community has been receiving benefits to the village fund through ticket sales for the rainforest canopy walkway (see Figure 5.3). Several of the Falealupo Chiefs[4] explained that ticket sales are distributed to the community under a split as follows: 10 per cent to the individual collecting ticket revenue, 45 per cent to his/her family and 45 per cent to the village fund. The Chiefs estimated that 800–1,000 WST ($342–$428) in fees are collected in average weeks in the high season

Table 5.1 Total monetary and non-monetary benefits related to the Samoan Mamala case

Monetary (USD)	Non-monetary
$85,000 to pay for the school in return for agreed access for R&D (Falealupo Covenant).	Conservation of the rainforest area in Falealupo.
Over $100,000 for construction and maintenance of a rainforest walkway.[5]	Community benefits including the schools and health care clinics.
Several other charitable donations to the community from Seacology and Dr Cox (approximately another $300,000).	Social recognition for the healers who have provided the knowledge about mamala.
Up to $70,000 in milestone payments under the ARA agreement if clinical trials are passed (ARA Agreement).	Opportunity for Samoan farmers to provide extract from the mamala plant (employment in biotrade activities) for further testing (although this case study could identify no evidence of this occurring in practice – synthetic analogs of prostratin appear to have been used recently in the United States instead of sourcing from Samoa).
6.7% royalties of revenues to Falealupo (net of expenses) if the ARA is able to license and commercialise a drug. Total returns to Samoa will be 20% of profits. 16.6% royalties of revenues to Falealupo (net of expenses) if UC Berkeley is able to license and commercialise a drug of a total of 50% of royalties to be returned to Samoa.	

and weekly collections are more like 200 WST ($85) at the bottom of the low season. Some interviewees noted that the village now depends upon the income from the walkway, and that this had caused some problems in the community. It was claimed that some have misspent some of the money and some of the chiefs operate the ticket booth more often than others, raising questions of inequity.

Other benefits do not appear to have arisen directly from the R&D, with the exception of an upfront milestone payment by the ARA to the Government of Samoa.

The other benefit that was raised by several interviewees was the payments made by the perpetual fund to the village council. This alleviates the need for individuals to pay into the village fund to pay for basic public works and maintenance for the village. This appeared to have continued for several years until recently when the fund was affected by the global financial crisis.

Discussion

Research on anti-retroviral applications of prostratin has not yet passed phase 1 clinical trials in the United States. Pre-clinical studies are still being

Figure 5.3 The rainforest Canopy Walkway Ecotourism attraction in Falealupo

Source: Robinson, 15/3/2012.

conducted by ARA. Given that a patent relating to prostration was filed in 1996, it appears likely that this patent will lapse prior to commercialisation of an anti-viral drug. Recently, the ARA have filed additional patent applications in 2009 on 'Methods of administering Prostratin and Structural Analogs Thereof' (US Application no. 12/937364; EPO Application No. 09730430.7; ARA, accessed 5/6/12). As Dr Cox and associates have noted:

> Synthesis of analogs . . . raises interesting issues concerning indigenous intellectual property rights. Because knowledge of prostratin's antiviral activity originated from ethnobotanical studies with Samoan healers, the AIDS Research Alliance (ARA) and the Government of Samoa agreed that 20 per cent of ARA's profit from prostratin will be returned to the Samoan people. Similarly, Samoa and the University of California, Berkeley, agreed to share equally in commercialization of the prostratin gene sequences. In the spirit of these previous agreements, we encourage future

developers of prostratin analogs for antiviral therapy to negotiate fair and equitable benefits with the Samoan people.

(Cox et al., 2008, p. 1589)

It is yet to be seen if the future users of synthetics analogs of prostratin will share benefits upon commercialisation of any drugs. Several of the commentary letters in the *Samoa Observer* are critical of exactly this point – that expectations for Samoa have been raised when it has become increasingly unlikely that benefits will arrive (Palamo, accessed 16/5/2012; Anonymous, 28/2/2002).

Although there have been few benefits provided to Falealupo directly 'arising from' the utilisation of genetic resources and traditional knowledge (taking a narrow view of the wording of the CBD and Nagoya Protocol), there have been considerable monetary and non-monetary contributions made to the conservation of the Falealupo rainforest, and to the Falealupo community. These reflect an acknowledgement of the contributions made by traditional knowledge holders to the potential development of a useful medicine, and the provision of access to genetic resources in the Falealupo rainforest. The establishment of a rainforest preserve directly responds to the first objective of the CBD, with caveats for traditional and sustainable use activities. Also, the related charitable contribution of the rainforest walkway provides a perpetual income for the village to be utilised for community activities, projects or infrastructure.

While some in the Samoan press have been critical of the ARA and UC Berkeley agreements, the royalty rates are quite high in comparison to many other negotiated figures for use of natural products in drug development. One issue that some interviewees raised was that expectations of short-term royalties were quite high following a visit from Dr Cox in approximately 2001, when he explained the ARA agreement (Fuiono Patolo, interview, 23/5/12; Seaumantufa Falemai, interview, 23/5/12; Manu Toifotino, interview, 24/5/12; Lemau Seaumantufa, interview, 24/5/12; Taii Tulei, interview, 24/5/12). Subsequently, interviewees and Chiefs indicated they had not heard about recent progress by the ARA and other researchers.[6] This highlights that it is important that the risks involved and long timelines for pharmaceutical R&D are clearly communicated to potential beneficiary communities with regular updates on progress.

Some other members of the public have also highlighted that prostratin was identified by researchers in New Zealand, some time prior to 1986. However, it does not appear to have been screened for anti-viral qualities until sent to the NCI and then subsequently patented for this intended use. Others have pointed out that the *Homolanthus nutans* tree is found across the South Pacific from approximately New Caledonia to French Polynesia (Whistler, 1996, 1994). Article 11 of the Nagoya Protocol highlights that in such circumstances, those Party to the Protocol should 'endeavour to cooperate, as appropriate, with the involvement of indigenous and local communities concerned, where applicable, with a view to implementing this Protocol'. The extent to which cooperation and transboundary benefit-sharing will occur is likely to be something that the Parties will resolve among themselves, through regional agreements, or through further negotiations at the Intergovernmental Committees of the Nagoya Protocol (ICNP).

It is not clear if traditional knowledge of Mamala is widely distributed across the Pacific. Whistler (1994) on Polynesian herbal medicine only described its use in Samoa, noting its use as a purgative for the digestive tract (also noted by Cox as a typical use, as well as by interviewees from Falealupo), as well as its use for bandaging circumcision wounds and for treating urinary tract infections (and rarely used in Samoa for viral infections) (Whistler, 1996). This may explain why Dr Cox undertook to share benefits with specific knowledge-providers and not others.

The case study also raises an interesting question about requirements for benefit-sharing relating to R&D towards synthetic analogs that are based on a naturally occurring compound. The Nagoya Protocol definition of 'utilisation of genetic resources' includes derivatives, meaning 'naturally occurring bio-chemical compounds'. Because analogs are synthetically produced, they probably do not fall under the scope of the Nagoya Protocol. However, this does not preclude Governments from specifying in benefit-sharing agreements that synthetic analogs utilised by the researchers involved must also share benefits (see Article 5.1 of the Nagoya Protocol on 'subsequent applications'). Enforcing third party benefit-sharing upon the development of synthetic analogs would be complex.

From interviews and comments in the *Samoa Observer*, it also seems clear that there have been conflicts over the sharing of money from the walkway, and over the agreements, even between Matai within Falelupo, and between upper and lower Falealupo. Without naming specific people, there was the suggestion that some people had benefited more through control of the revenue from the perpetual fund and from the walkway than others, and that decisions had been dominated by some parts of the community who liaised with Cox. As an outside observer it is important not to judge or to expect to fully understand these local (and national) matters because they are embroiled in local culture, custom and politics. Rather it is an important reminder that in 'making the cut' and including some to the exclusion of other people or groups in the negotiations and benefits, there is inevitably going to be some tension or conflict. It seems that it may be a 'Western' expectation that for the sake of a legal agreement (inscribed as text, rather than orally agreed and locally/socially mediated as is often the case in traditional customary laws) a 'community' of beneficiaries must be clearly defined. This process might be perceived as paternalistic or imposed in-country, and is not isolated to this case – it is also discussed by Strathern (1999, as counterposed to the social relations embedded in gift-giving amongst Hagen men in Papua New Guinea) and Hayden (2003, on 'idioms of inclusion' in the ICBG Latin America), among others. As Whatmore (2002, p. 103) noted:

... the boundaries between [plant genetic resources (PGR)] as a gift of nature or a social artefact return the disputations of this collective event[7] again and again to the alignments of sovereignty and property and the legal practices that affect them.

Thus it is the same here in this case, that despite many benefits for the local community in Falealupo (which have nevertheless become imbricated in local tensions) these same questions about ownership, sovereignty and property have arisen. This has then driven the discussion to questions about 'fairness and equity' – a critical intension of the ABS aspects of the CBD and Nagoya Protocol – between the US researchers and Samoa, and also between the relevant actors within Samoa.

Notes

1 This was incorrectly understood by the healers or translated by Dr Cox – the treatment was in fact for hepatitis as evidenced by jaundiced yellow skin.
2 Fuiono Aleki, Taii Tapana, Tapua Tamasi, Manutuaifo, Kelemete, Gaga Sanele, Ulufanua Aleuna, Kolone Va'ai (pers. comm. 15/3/12) and Solia Papu Va'ai (pers. comm. 13/3/12).
3 While most were happy with this, one interviewee noted that these terms were paternalistic.
4 Fuiono Aleki, Taii Tapana, Tapua Tamasi, Manutuaifo, Kelemete, Gaga Sanele, Ulufanua Aleuna, Kolone Va'ai (pers. comm. 15/3/12).
5 Although this is a philanthropic contribution, rather than a benefit arising from access to genetic resources, it has been included here because it demonstrates a subsequent commitment to the community post-access and during R&D.
6 Fuiono Aleki, Taii Tapana, Tapua Tamasi, Manutuaifo, Kelemete, Gaga Sanele, Ulufanua Aleuna, Kolone Va'ai (pers. comm. 15/3/12) and Fuiono Patolo, Seumanu Tafa Faaolo, Seumanu Tafa Falemai, Manu Toifotino, Lemau Seaumantafa, Taii Tulei, Marianive Fuiono (interviews, 23–24/5/12).
7 She is referring specifically to collection of germplasm for ex situ collections.

References

Anonymous, 'The strange story of the Samoan "AIDS tree" '. *Samoa Observer,* accessed 28/2/2002.
ARA. '*Our Progress in Developing Prostratin*'. Available at: http://aidsresearch.org/cure-research/our-progress, accessed 5/6/12.
ARA. '*Prostratin*'. Available at: http://aidsresearch.org/cure-research/prostratin/, accessed 21/1/2014.
Cox, P. (2001) 'Ensuring equitable benefits: The Falealupo covenant and the isolation of anti-viral drug prostratin from a Samoan Medicinal plant'. *Pharmaceutical Biology,* 39(supplement), 33–40.
Cox, P., Johnson, H.E. and Tavana, G. (2008) 'Giving Samoan healers credit for prostratin'. *Science,* 320(5883), 1589.
Hayden, C. (2003) 'From market to market: Bioprospecting's idioms of inclusion'. *American Ethnologist,* 30(3), 359–371.
Palamo, A. ' "Who really owns it?" and "who really owns it part II and III"'. *Samoa Observer,* accessed 16/5/2012.

Strathern, M. (1999) *Property, Substance and Effect: Anthropological Essays on Persons and Things.* Athlone press, London.

Whatmore, S. (2002) *Hybrid Geographies: Natures, Cultures, Spaces.* SAGE, London.

Whistler, W.A. (1994) *Polynesian Herbal Medicine.* Pacific Tropical Botanical, Hawaii.

Whistler, W.A. (1996) *Samoan Herbal Medicine.* O le Siosiomaga Society Inc of Western Samoa.

Whistler, W.A. (2004) *Plants in Samoan Culture: The Ethnobotany of Samoa.* Isle Botanica, Honolulu.

6 The Moroccan Argan case

This chapter examines the use of argan tree (*Arganiaspinosa (L.) Skeels* – see Figure 6.1) products in the cosmetics, skin and hair care industry, with a particular focus on the 'corporate social responsibility' (CSR) approach of BASF's personal care business (at the time of first contact Laboratoire Sero-biologiques was the partner – a division of the company Cognis, which was acquired by BASF in 2010), working with L'Oreal, the NGO Yamana and their partners in Morocco – the Targanine cooperatives. In this particular case, it is worth noting that Morocco had no ABS laws or regulations at the time of the arrangement. Also, taking a literal view of the CBD text, some

Figure 6.1 An argan tree in the Souss Valley, near Tioute, Morocco.

Source: Robinson, 4/8/2010.

from the industry have suggested that companies only needed to establish ABS agreements where 'genetic resources' or genes containing functional units of heredity were utilised for research and development (R&D) (although others have suggested that there was an implicit obligation regarding benefit-sharing for biochemical 'derivatives' of genetic resources since the Bonn Guidelines were established). In this case, the companies have described their arrangement as a 'CSR' approach and the project has many benefits, including a 'social fund' which acts much like a benefit-sharing fund for utilisation of a biochemical fractions from extracts derived from the argan tree.

In any case this arrangement between producers and suppliers is of relevance in the context of the new international ABS developments. Representatives of these companies have emphasised that they are dedicated to continued learning and improvement regarding this agreement, for the benefit of the Berber women involved (Barre, 26/7/2010; D'Erceville, 26/7/2010) – the 'providers.' Therefore this case presents a hypothetical of the sort of arrangements that would take place under the new internationally binding requirements under the Nagoya Protocol on ABS, once national systems have been established. It was in this context that the author first contacted Cognis (in 2010) to find out more about their CSR approach to argan products, in order to identify synergies between fair trade, sustainable use and ABS approaches for industry.

Figure 6.2 Argan fruit drying in the sun in front of an argan tree-covered mountain, Northern Souss Valley, Morocco.

Source: Robinson, Approximately 28/4/2011.

The argan tree is endemic to Morocco and provides important ecological functions with deep root systems allowing for survival and stabilisation of soils in the arid ecosystem of the Souss Valley of southwest Morocco, the Anti-Atlas mountains and surrounds bordering the Sahara desert (Figure 6.2). For the Amazigh or Berber peoples of the region, the argan forest provides an important income source through the sale of oil for culinary, cosmetic and medicinal purposes, firewood and charcoal for heating and cooking and food and shade for livestock (especially goats) (Lybbert et al., 2002). In recognition of its ecological value and socio-economic importance, the argan forest region was declared a United Nations Educational, Scientific and Cultural Organization (UNESCO) biosphere reserve in 1998.

Argan oil has been used by the Amazigh for hundreds, if not thousands of years, as a food oil and as a skin or hair care product (Charrouf and Guillame, 2009). The fruits of the argan tree are collected once they have fallen to the ground, dried in the sun, and de-pulped. Once de-pulped a hard nut is manually broken by the women, usually by striking it with a flat stone on another hard surface, to reveal an almond-like kernel. The Amazigh women[1] traditionally pressed the kernels by hand to extract the oil using stone implements, but this part of the process has since been mechanised. However, the cracking of the hard nuts is not yet mechanised, despite attempts by some people to devise and patent a machine for this purpose. During interviews, the women explained that if the kernel is shattered or broken then it must be discarded as it will affect the quality of the oil, and that most attempts to mechanise this aspect have failed to keep the kernels intact. This means that the labour of many Amazigh women across the Souss Valley is still required for the oil production, and provides an income for many households in the region.

The oil of pressed argan kernels is now known to be high in vitamin E, carotenes and essential fatty acids, making it ideal for use in skin and hair care products. The first written record known to exist regarding these uses is from the famous physician Ibn Al Baytar in 1219 (Charrouf, interview, 4/8/10). The market for argan products has recently expanded, servicing the local and tourist populations, as well as a significant export market. As Lybbert (2007) noted 'Introducing cosmetic argan oil into high value international markets has required research into both the chemical properties of the oil and potential extraction and processing technologies'. This need to conduct R&D raises the relevance of ABS and has even led to some criticisms about the apparent lack of ABS agreements made by companies in the 1990s and early 2000s (McGown, 2006). Between 1994 and 2005, Prof. Zoubida Charrouf helped a group of Amazigh women to establish the 6 cooperatives (Ajdigue, Tagmate, Tamaynoute, Targante, Toudarte and Taitmatine) that now operate under the economic interest group (EIG) and brand 'Targanine', with the aim of benefiting and empowering local women for supply of Argan oil and products to these burgeoning markets. These Targanine cooperatives, through their connections with Prof. Charrouf, then entered into commercial agreements with Cognis and subsequently L'Oreal that are of interest in the ABS context.

For this case study, a number of key stakeholders were met and interviewed, including meetings with staff of BASF, L'Oreal and Yamana (26/7/2010, 4–5/5/2011), interviews and meetings with Targanine staff (26, 28, 29/4/2011), an interview with Prof. Zoubida Charrouf (4/8/2010), the managers of 6 Targanine cooperatives and several informal focus group-style interviews with groups of women while they were cracking argan nuts at the Targanine cooperatives and their centres (26–29/4/2011, and at Taitmatine on 4 /8/2010).[2]

Access and R&D conducted

The traditional use has prompted further investigation by researchers and companies for decades now, with significant research activity beginning in the 1980s with firms focusing on culinary oil; pure Argan oil for cosmetic uses; and also R&D into extracts of bioactive ingredients from Argan oil, leaves, fruit and seeds – this is where ABS is relevant (Lybbert, 2007). In this case we are most interested in the research established as a partnership between Prof. Zoubida Charrouf and Cognis around 2000, with patent applications occurring in the United States and Europe from 2001, with patents granted in 2005, 2006, 2008 and 2011. These include:

- Cosmetic and/or dermopharmaceutical preparations containing native proteins from the plant *Arganiaspinosa* (US Patent Number 7,871,766, granted on 18 January 2011, first filed on 28 November 2001);
- Cosmetic and/or dermopharmaceutical preparations containing leaf extracts of the plant *Arganiaspinosa* (US Patent Number 7,105,184, granted 12 September 2006, first filed 28 November 2001);
- Cosmetic and/or pharmaceutical preparations that contain an extract of the plant *Arganiaspinosa*. (2000) (EP1276460B1);
- Claim 1: A cosmetic and/or pharmaceutical preparation that contains saponins from an extract of the plant *Arganiaspinosa*;
- Claim 2. The preparation as claimed in claim 1, characterised in that the extract is obtained by extraction from parts of plants, selected from the group consisting of the leaves, the roots, the stem, the bark, the flowers, the fruits, the fruit flesh and the seeds;
- A plant extract and its pharmaceutical and cosmetic use. (2003) (EP1711194B1);
- Claim 1: The triterpene-fraction of the extract of the pulp of the fruit of *Arganiaspinosa*;
- Claim 2. The non-saponifiable-fraction of the extract of the pulp of the fruit of *Argoniaspinosa*.

L'Oreal also has filed several patent applications which note the use of argan oil as a potential ingredient in vegetable oils used for skin or hair care products. They also cite the use of BASF branded extracts such as Arganyltm (leaf extract) in some of their patented innovations. Notably, none of the patents appear to make claims over the oil or extracts from the oil. This is likely due to the existing prior art and traditional knowledge of the oil for cosmetic purposes, and due to the need for 'inventiveness' or 'non-obviousness' as a patent criteria. It is likely that a biochemical extract from the oil, which is known to have cosmetic benefits, would be obvious to a dermatologist – someone 'skilled in the art/science'. Interestingly, the Cognis patents thus focus on proteins extracted from a by-product, and also flavonoids isolated from the leaf. Arguably, these are considerably more 'novel and inventive' biochemical isolations than if they had been focused on argan oil.

Because Morocco does not yet have a designated ABS law, there was not an explicit 'access' requirement in the country for 'utilisation of genetic resources'. During meetings with Cognis and L'Oreal staff, it was emphasised that the required legal permissions, such as phytosanitary certificates, had been obtained. Preliminary research was also conducted in-country by Prof. Charrouf (who is named on a number of the patents as an inventor) (Charrouf and Guillaume, 1998), prior to further R&D by BASF and L'Oreal towards a number of marketable skin care and hair care products. Some of these products utilise commercial argan oil (outside of the scope of ABS), while other products utilise the specific biochemical extracts derived from the argan tree that have been identified through R&D (this would potentially be within the scope of the Nagoya Protocol concept of 'utilisation of genetic resources' once it comes into force and is nationally implemented by countries). Because this R&D occurred prior to the Nagoya Protocol and without domestic ABS laws in Morocco, this case can then be considered an ABS hypothetical.

Benefit-sharing

A tripartite partnership has been established since 2008 between BASF, L'Oreal and Yamana for the implementation of a CSR program along the supply chain for supply and use of argan products. The Targanine cooperatives in Morocco supply argan oil and related argan products (e.g. pressed oil cake and argan leaves) to BASF under fair trade arrangements of prepayment for the product (2 year contract for the supply of oil), pre-agreed prices and a premium price paid (e.g. 5 per cent premium paid for oil at the end of the year since 2010). BASF maintains elements of quality control and refinement of the raw products. L'Oreal is supplied with the refined product which it then incorporates into a number of cosmetics and related skin/hair care products that are distributed internationally, making the whole partnership economically viable. Yamana has a key role as a trainer and facilitator working with the cooperatives in Morocco, ensuring that local stakeholder expectations are being taken into consideration, that there is traceability in the supply chain and payments through cooperatives, and to incorporate value addition into the supply chain (Robinson and Defrenne, 2011). The traceability is an important element, allowing the EIG to ensure that the appropriate amount of money is getting to the producers (the women in the cooperatives) and that it is not being taken by middlemen or others in the supply chain. Traceability and bookkeeping was also emphasised for determining where argan fruit was collected, such that it qualifies for organic certification, as well as geographical indications protection designating that the oil is from the Souss Valley.

While there are several notable elements in the partnership that benefit the Amazigh women through the Targanine cooperatives (discussed further in Robinson and Defrenne, 2011), the most relevant aspect to this ABS example is regarding the use of argan 'pressed cake'. The pressed cake is essentially a by-product of the argan oil extraction process, from which useful protein fractions have now been identified (as described in US Patent Number 7,871,766) (see Figure 6.3). The cooperatives sell pressed cake to BASF for approximately 15 times local market prices. Indeed interviews at the Targanine cooperatives revealed that the by-product is of little use to locals, except perhaps as feed for goats. This substantial 'premium' price paid for the by-product

Figure 6.3 Bowls of argan product illustrating the process: dried fruit, nut, cracked nut, almond/kernel, pressed cake (by-product), and bread used to dip into edible argan oil.

Source: Daniel Robinson, approximately 28/4/2011.

might be thought of as a fair trade approach to ABS, or as a way of linking ABS with 'biotrade' (see Oliva, 2013). This means tens of thousands of Euros are annually paid to the cooperatives for tonnes of the pressed cake by-product,[3] even though it has a much lower local market value. The resulting payment is split such that:

- Approximately 50 per cent goes to a social fund for each cooperative;
- Approximately 25 per cent to the EIG of Targanine (the formal business and marketing operation of the Targanine cooperatives) for maintenance costs, machinery, investment and management; and
- Approximately 25 per cent to the cooperative (indirectly for incomes) to be spent or divided as decided by general meetings of the shareholders of the individual women's cooperatives.

The social fund of each cooperative that receives 50 per cent of this premium payment for pressed cake is quite similar to other benefit-sharing funds from ABS agreements. Each cooperative decides how to spend the money within the fund and so it has been allocated to a number of different purposes. Of the 25 per cent to the EIG, in 2011 (from the 2010 payment) the money was allocated towards logistical costs (transport from the cooperatives to the EIG), the local quality control costs, the packaging, paperwork costs for exportation and a profit margin for EIG. The purchase of a van to take guests (such as the Yamana in-country facilitator and myself), auditors and potential buyers to the cooperatives in 2010 was from the 5 per cent fair trade premium paid for the oil at the end of the previous year. This decision was again made by consensus at an annual general meeting of the Targanine EIG.

Further to this, the leaves are used to derive an extract at BASF, which has a cosmetic application. The leaves are collected on parcels in the argan forest under a covenant with the Moroccan Water and Forests Authorities. The covenant specifically implies commitments for maintenance of trees and reforestation of the parcel. Harvested leaves could be considered to be by-products of trees maintenance operations. In a meeting with staff from the High Commission (Regional Office for the South West), they indicated that they had been reluctant to allow or licence groups to cut leaves from the trees, unless under strict conditions. They indicated that it might cause an enforcement problem if not carefully managed (discussed further below), and that they had to prevent damage to trees in the UNESCO Biosphere Reserve – the 'Arganosphere' as it is sometimes called. The 'Haut Commissariat aux Eaux et Forests' (High Commission for Water and Forests) regulates the collection of argan fruit and upholds usufruct rules surrounding its collection. People are only allowed to collect the fruit from the ground, and goats are restricted from climbing and eating the fruits during certain months. These rules have had to be applied more strictly in recent years due to the booming market for argan oil, meaning that people can be fined for collecting from trees or damaging them.

Impact

The social fund has been spent on a range of different expenses, as decided by each of the 6 cooperatives. These include the following:

- Literacy programs (Arabic);
- Health insurance fund which is only eligible for under 65 year olds (notably, some cooperatives had not opted for health insurance so as to ensure equity among all ages of women in their cooperative);
- Medical expenses for the women and families (and the cost of some surgeries);
- School supplies for children (books, pens, furniture);
- Distribution of food among the local community;
- Weddings and supplies for weddings;
- Optometry testing and the purchase of eye glasses;
- Cushions and air conditioning for cooperative rooms to improve comfortability;
- Pharmacy products;
- Washing machines (shared at the cooperative);
- Sewing machines and classes;
- Playground, toys and games for children;
- Blankets for families for the wintertime; and
- TVs and whitegoods.

During interviews with the women, they commented on the many benefits that had arisen from the tripartite partnership, and particularly, from the social fund. These included the following comments in response to the questions (in bold):

- **Before joining what did you do?** 'We didn't do anything much – we were just at home'.
- 'We did prepare some argan oil for ourselves, housework, sleeping'.
- 'Looking after kids. We had no money'.
- **Now how has your life changed?** 'We have our *own* money'.
- 'We can read and write Arabic'.
- **What do you do with your money:** 'Anything we want (laugh)! Go to Agadir, buy clothes for us and the children. Furniture, white goods (fridges)'.
- **Where are the kids:** 'They come here. Some are at home with their dads, some at school'.
- **Do you like it here:** 'Yes. We are like a family. We talk and eat together here. There is solidarity. We respect each other (women) and their time'.
- 'We feel very strong and independent. We can open our own purse and have our own money'.
- **The Social Fund:** 'We have changed so many things, done so much with the money'.

- 'We have helped our families, the local community and ourselves. Next we will go to Mecca (joking – they all laugh)'.
- **Do you know where the oil goes:** 'Yes. People from L'Oreal came once and gave us samples of the cosmetics'.

The impacts of the social funds were quite noticeable in some cooperatives where the money had been spent on equipment or a crèche for the children (see Figures 6.4 and 6.5). In other cooperatives, the women had bought health insurance but nothing else because they could not decide how to spend the money. They explained that they were not used to having money to spend.

The women and the interviewees in the partnership emphasised the importance of not just the social fund, but also the range of other improvements and benefits that were felt as a result of the partnership. For example, some of the women explained that they now earned more than their husbands and that their earnings from the cooperative were a primary source of income. When I asked if this had been a source of household tension, several women at different cooperatives explained that they would only be allowed to join the cooperatives if their husbands were supportive and their family wanted or needed the income. In one of the discussions, a woman noted that there

Figure 6.4 Sewing machines purchased with the social fund at cooperative Toudarte. Behind the machines are school-books for the children, also from the social fund.

Source: Robinson, 29/4/2011.

Figure 6.5 Playground equipment at Toudarte purchased by the social fund so that the
women can work and know their children are playing safely nearby before and/
or after school.

Source: Robinson, 29/4/2011.

had been some marital tension and that the proliferation of argan cooperatives
might lead to more divorces. It was evident that certain communities were
more conservative than others and so there were differences in perspective
between the cooperatives about the impacts of the industry on family life (see
also le Polain de Waroux and Lambin, 2013).

At all of the cooperatives we asked what an average daily income would
be for one of the women. This always led to detailed discussions about the
amount paid and the number of hours worked, the time it took different
women to crush the nuts and also how much fruit that woman and her family had
collected. Women who are part of Targanine are typically paid per kilogram
of almond produced (by cracking nuts), and they are also paid per kilogram of
fruit collected and supplied to the cooperatives. At cooperative Toudarte they
had clearly worked out an average daily figure of 75–85 Dirham ($10) per day
for a woman producing 700 g to 1 kg of almonds (a typical amount for a day's
work, or approximately 7–8 hours) (figures based on interviews in 2011, and
correspondence with BASF in 2014). This figure might be even higher if the
woman supplies fruit – paid at approximately 1 Dirham per kg. Similar figures
were estimated at the other 6 cooperatives, or slightly lower daily rates (45
Dirham per day for a woman who was a slow worker, and 65–85 Dirham per

day for women who are efficient). Eighty five Dirham per day is significantly above the agricultural minimum wage of 63 Dirham (approximately $7) per day (up to 10 hours), and the women have more flexible and typically shorter working hours. This is without the additional income supplement provided by the social fund, from which the women can vote to receive a portion of this money for specific purposes (e.g. purchase of white goods), or the redistribution of yearly benefits (from the fair trade allocation). Further to this, the members of the partnership such as EIG Targanine and Yamana have assisted with applications to the European Union for grants for the purchase of equipment for pressing the almonds, quality control and also for de-pulping the fruit (also co-financed by cooperative funds, including the social fund). This equipment results in a higher value product for sale to BASF or to other buyers (they have a non-exclusive agreement). It also means that the women are able to save time – for example, it might take a woman an entire day to manually de-pulp the outside fruit layer from one 60 kg sack of argan fruit. There is no income derived from de-pulping – it merely prepares the nuts for cracking. However, the purchase of de-pulping machines reduces this time to remove the dried fruit husk to about 5–10 minutes. By purchasing the machines (a few of the cooperatives had them as of August 2011), the women can effectively double their almond production time and income.

Further to this, the success of the cooperatives has meant other monetary benefits for them. To become a member of the cooperative required the women to make an upfront payment of about 400 Dirham in 2004 at Toudarte.[4] This 'share' in the cooperative is worth about 15,000 Dirham ($1800, in 2011), so when women retire from this work they can draw back their share.

Discussion

There are clearly considerable benefits for the women in the Targanine cooperatives from the purchase of argan oil by BASF and other European companies (biotrade) and also from the social fund for the purchase of pressed cake (from which a type of benefit-sharing occurs through a premium price paid). If we examine the Annex of the Nagoya Protocol, the list of monetary benefits includes access fees/fees per sample, up-front payments and special fees to be paid into trust funds supporting conservation and sustainable use of biodiversity, salaries and preferential terms where mutually agreed, among other things. Often benefit-sharing involves access or up-front fees, then milestone payments and/or royalties at the achievement of specified developmental milestones. While this case is different, the premium payment for commodity supply of argan pressed cake into a trust fund is quite similar to the suggested 'special fees' item in the annex. In any case, the Annex to the Protocol is a non-exclusive list and premium price for biotrade supply could well be an addition. A number of non-monetary benefits such as contributions to the local economy, social recognition and food and livelihood security benefits are also relevant to the cooperatives. Other items from the Annex such as joint IP and collaboration in research, participation in product development and also technology transfer are also relevant given the collaborations between Prof. Charrouf and BASF, as well as the technology transfers made to the cooperatives.

The benefits are felt by the women who have a share in the cooperative (e.g. at Tagmate there are approximately 40 women), with flow on benefits for their community if the women choose to spend their money in certain ways (e.g. setting up medical treatment days or eye-sight testing for the community). However, the argan trade has been so lucrative, particularly for the Targanine cooperatives that benefit from the social fund, so that now many other women wish to join, but are not able. Given that there is such demand for argan oil by local and international markets, including the bulk purchases by BASF, the cooperatives are having trouble keeping up with demand. At the time of interview in 2011 they sometimes had to purchase fruit from families and brokers in surrounding areas because the women in the cooperatives cannot supply enough on their own. In addition, the cooperatives have attracted affiliated centres whereby the Targanine cooperatives had to purchase almonds from these women at standard market rates (depending upon the decisions made by that cooperative) in 2011. This has since been improved with agreements implemented between cooperatives and affiliated centres. These agreements include commitments on the price paid, an annual premium for the affiliated centre (fair trade) and access to the social fund actions.

This means that the dissemination of fair trade payments and social fund benefits is limited to a relatively small group (approximately 240 women in all 6 cooperatives as of August 2011), and this might raise questions about 'fairness and equity'. This rings true of the industry more broadly with the findings of Lybbert et al. (2002, p. 125) who noted that most locals 'only participate superficially in the new and expanded markets for argan oil, and the benefits that do trickle down to local households appear to be regressively distributed, both regionally and between households'.

While the cooperatives have continually expanded, the EIG explained that they can only do so at a relatively slow rate because of the risks associated with market volatility, demand and also the quantity they can obtain from the argan forest. In the ABS context this means that the local 'providers' of the 'genetic resource' in the Nagoya Protocol sense are benefiting in the absence of a broader ABS framework in Morocco for, say an ABS fund. In other circumstances ABS funds have been established with the intent to disseminate benefits to a wide range of beneficiaries – in this case the traditional knowledge about the cosmetic benefits of argan is widely held and so there might be a claim for wider benefit-sharing. However, the patented pressed cake extract is arguably *not* directly based on traditional knowledge because it is from a processed extract not previously used by the women. This case also differs from others because the provision of genetic resources occurs as a biotrade transaction – tonnes of pressed cake purchased by BASF and used by L'Oreal. The original access to genetic resources for utilisation (R&D) was arguably made by Prof. Charrouf, the Moroccan scientist, who then began to work with BASF and

L'Oreal towards unique biochemical extracts as active ingredients for cosmetics. As a co-inventor, Prof. Charrouf has received some benefits – although she indicated that the direct monetary benefits she has received have been limited. Effectively, by helping establish the EIG Targanine group, Prof. Charrouf has ceded most benefits to those Amazigh women shareholders of the cooperatives. The women continue to benefit (rather than a one-off payment, as is common for 'milestones' in commercial development of medicines in ABS agreements) because of the biotrade supply chain of both the oil and pressed cake (and to a lesser extent the argan leaves where the EIG Targanine is an intermediary in the supply chain). In this sense the agreement and supply chain is probably more successful at delivering benefits over a longer time period and to a significant number of beneficiaries than many other ABS-type agreements, precisely because it is premised on continued premium price paid for a biotrade supply of product. So, despite questions of distributional equity, I would argue that the arrangement provides significant and life-changing benefits to a considerable number of women and their families. If this was instead put into a centralised fund to be distributed more broadly in Morocco or even the Souss Valley region, the funds might be so diluted as to be less effective.

Although we didn't have the opportunity to interview many women outside of the cooperatives to gain external perspectives, we were able to interview some women in the centres. Several of these women expressed that they were happy just to have work. One of the managers of a centre indicated that all of the women in her centre would like to formally be part of the EIG Targanine to receive the benefits – however, she also expressed her happiness to have the paid work. At other non-Targanine fair trade argan oil producing cooperatives in the Souss Valley near Agadir that were interviewed at a later date (in 2012), some of the managers knew of Targanine but did not know of the explicit benefits received or differences between their cooperatives.

In terms of the broader impact of the activity, argan oil production is often described as a 'sustainable use' activity because the fruit is collected from the ground and so should not necessarily affect the health of the individual trees, or the conservation of the forest as a whole. As Lybbert et al. (2011, p. 13963) noted, 'Win–win claims of poverty alleviation and biodiversity conservation now appear on virtually every argan product label and have been showcased by media outlets worldwide'. The cooperatives only received a limited amount of training and education about the conservation of the forest, apart from basic information about the regulations surrounding collection (which in any case relate to their customary rules surrounding argan trees). Rather, training by Yamana often focused on quality control, traceability and value addition. At the specific cooperatives that we visited, the women all knew that they were not allowed to cut or damage trees. However, some authors have raised concerns about the sustainability of the activity. First, the over-collection of fruit from the ground can affect the extent of natural tree germination. Argan tree growth is slow, but the tree can be extremely long-lived to 125–150 years, and the alkaline soils of the region may affect seed germination (Morton and Voss,

1987), suggesting a slow replacement rate. From conducting household surveys in the region, Lybbert et al. (2011) noted that conflict over argan resources has increased, including conflicts over the erection of permanent barriers around argan trees that would normally be open for seasonal usufruct collection and grazing by goats. They noted that locals tend to use sticks to dislodge argan fruit and harvest more aggressively, damaging the trees and buds for subsequent year's production (Lybbert et al., 2011). The authors warn that forest exploitation is shifting towards short-term fruit collection rather than long-term concerns about forest sustainability. Other authors have noted that there has been a decrease in forest density over the past decade, undermining production (le Polain de Waroux and Lambin, 2012). They also suggest that there are some potential perverse impacts from the booming industry. First, the increase of revenue into some households is sometimes spent on expanding goat herds – the expanding goat herds feed in the Argan forest and may be having a long-term impact on it. However, some research on household livelihoods in the region suggests that income related to argan oil marketing has been significant for only some, while being limited compared with non-farm work and remittances (le Polain de Waroux and Lambin, 2013). Second, Lybbert et al. (2011) noted from household surveys that girls who stand to benefit from the argan boom were significantly more likely to make the transition from primary school to secondary school. This is an interesting finding because a couple of our interviewees in the cooperatives and their affiliated centres noted the potential perverse incentive of the argan boom may be that some materially poor families will ask their daughters to crush argan nuts for income rather than sending them to school. Given that the argan industry often touts itself as benefiting and empowering women, this is something to be monitored by the cooperatives.

In summary, there is a significant effort made by this partnership to improve the quality of life of the women who work in these cooperatives. It is evident from the interviews that these benefits have made a positive impact upon them and their families. While there are some potential concerns and questions that can be raised about equity and sustainable use, it is worth acknowledging that gradual improvements have been made under this arrangement. It is important that industry is recognised and encouraged to continue this 'corporate socially responsible' behaviour so that they have the incentive to continue in this and other contexts. No ABS agreement is ever likely to be perfect, and I would argue that we should not be too disparaging of the potential issues that arise where genuine collaborative efforts are being made, as we see in this case.

Notes

1 The production of argan oil has traditionally been undertaken by Amazigh women, and this is still the case today.
2 Interviews were undertaken in this manner so that it allowed several different women from the group to give their perspectives on the Argan production process, without distracting them from their work.

3 Exact figures were provided by Targanine, but members of the tripartite partnership requested that this information not be publically disclosed for commercial-in-confidence reasons.
4 It should be noted that Toudarte is probably the most successful of the 6 cooperatives and so the shares may not be worth as much at the other cooperatives.

References

Charrouf, Z. and Guillaume, D. (1999) 'Ethnoeconomical, Ethnomedical, and phytochemical study of *Argania spinosa* (L.) skeels: A review'. *Journal of Ethnopharmacology*, 67(1), 7–14.

Charrouf, Z. and Guillaume, D. (2009) 'Sustainable development in Northern Africa: The argan forest case'. *Sustainability*, 1(14), 1012–1022.

le Polain de Waroux, Y. and Lambin, E.F. (2012) 'Monitoring degradation in arid and semi-arid forests and woodlands: The case of the argan woodlands (Morocco)'. *Applied Geography*, 32(2), 777–786.

le Polain de Waroux, Y. and Lambin, E.F. (2013) 'Niche commodities and rural poverty alleviation: Contextualizing the contribution of argan oil to rural livelihoods in Morocco'. *Annals of the Association of American Geographers*, 103(1), 89–607.

Lybbert, T.J. (2007) 'Patent disclosure requirements and benefit sharing: A counterfactual case of Morocco's argan oil'. *Ecological Economics*, 64(1), 12–18.

Lybbert, T.J., Aboudrare, A., Chaloud, D., Magnan, N. and Nash, M. (2011) 'Booming markets for Moroccan argan oil appear to benefit some rural households while threatening the endemic argan forest'. *Proceedings of the National Academy of Sciences*, 108(34), 13963–13968.

Lybbert, T.J., Barrett, C.B. and Narjisse, H. (2002) 'Market-based conservation and local benefits: The case of argan oil in Morocco'. *Ecological Economics*, 41(1), 125–144.

Morton, J.F. and Voss, G.L. (1987) 'The argan tree (*Arganiasideroxylon*, Sapotaceae): A desert source of edible oil'. *Economic Botany*, 41(2), 221–233.

Oliva, M.J. (2013) 'The Implications of the Nagoya Protocol for the ethical sourcing of biodiversity'. In Morgera, E., Buck, M. and Tsioumani, E. (eds) *The 2010 Nagoya Protocol on Access and Benefit-sharing in Perspective Implications for International Law and Implementation Challenges*. Martinus Nijhoff Publishers, Leiden, pp. 371–388.

Robinson, D.F. and Defrenne, E. (2011) *Argan: A Case Study on ABS*? Union for Ethical Biotrade, Geneva.

7 The Australian ABS system and examples

Having discussed a number of biodiscovery access and benefit-sharing (ABS) agreements or ABS-type examples from countries that have no ABS system (e.g. Morocco, Madagascar, and Samoa) or that have limited systems that are only partially implemented (e.g. Thailand, Vanuatu – see Chapter 8), it is worth also examining a country that has an existing ABS regulatory framework. The concept of ABS under the CBD and Nagoya Protocol rely on contracts agreed through mutually agreed terms (MAT), and so it has been possible for several existing ABS-type agreements to have been established under contract laws of those specific countries in the absence of a national system. In contrast, it is interesting to see what the experience of ABS has been in countries where regulatory systems are in place. For this particular discussion it is worth examining the 'Nationally consistent approach for access to and utilisation of Australia's native genetic and biochemical resources' and some of the laws and agreements that have been made, and permits issued. As a mega-diverse country, Australia has a significant number of unique and endemic species that have evolved over time on the island continent. Given the provision of 'sovereign rights' over biological diversity under the CBD, this means that Australia could stand to gain significantly from ABS arrangements if they can be clearly established. For example, approximately 83 per cent of mammals in Australia are endemic to the country (Laird et al., 2008). In addition, Australia's unique marine and coastal environments contain the world's largest areas and highest diversity of tropical and temperate seagrass species and of mangrove species, some of the largest areas of coral reefs, exceptional levels of biodiversity for a wide range of marine invertebrates, and it is estimated that around 80 per cent of the southern marine species are endemic (DFAT, 2008; Laird et al., 2008).

The 'nationally consistent approach' to ABS was endorsed in 2002 by the Natural Resource Management Ministerial Council of Australia, to establish a set of principles for each jurisdiction in Australia (the states and territories) to implement the ABS aspects of the CBD. The framework was based on the Bonn Guidelines and was developed such that states and territories agree to certain framework principles, while still retaining some flexibility in how they implement appropriate laws, policies and/or administrative measures.

Australia recently signed the Nagoya Protocol, and its Commonwealth Department of Environment is making consultations and preparations to ratify (DoE, 2002).

Under this system, certain states and territories have functioning ABS laws and systems. From these a number of ABS agreements have been made, and there are many non-commercial permits granted. Often for the non-commercial permits there are no benefit-sharing provisions except for data sharing. Many of the commercial agreements have even been small in scope and benefit. However, they highlight that benefit-sharing does not need to be grand in scale nor likely to generate significant milestone or royalty flows, especially where the odds of commercial success are particularly low or where the research is of an academic nature or poorly funded. This chapter focuses particularly on the 2 jurisdictions which have the most established regulations dealing with ABS: The Commonwealth and Northern Territory.

Access to biological resources in commonwealth areas

For areas under Commonwealth jurisdiction in Australia (e.g. certain national parks and defence land), the *Environmental Protection and Biodiversity Conservation Act* (EPBC) 1999 and Regulations specify ABS requirements. For example, these Commonwealth areas include land owned by Commonwealth agencies and departments, land areas leased by the Commonwealth and its agencies (such as areas on Norfolk Island, Kakadu National Park, and Uluru-Kata Tjuta National Park), and the continental shelf, the waters of the exclusive economic zone and the seabed under those waters and the airspace over them (Department of Environment, 2014). Specifically, the permitting and ABS rules are outlined in the EPBC Regulations, Part 8A. Under this framework, the government has sought to provide a streamlined permit process for non-commercial research (see Box 7.1), as opposed to commercial research (or potentially commercial research). The vast majority of applications to the Commonwealth Government are for non-commercial research access to biological resources (notably, the government uses this broader term, than the more specific 'genetic resources' used by the CBD). For those that apply for commercial access, an additional requirement is for the establishment of a benefit-sharing agreement with the 'access provider', which is typically the Protected Area Policy and Biodiscovery Section of the Commonwealth Department of Environment (DoE, formerly DSEWPAC).

> ### Box 7.1 Procedure for accessing biological resources from commonwealth areas
>
> When applying for a permit to access biological resources for non-commercial purposes in a Commonwealth area, an applicant must obtain written permission
>
> *(Continued)*

(Continued)

from each Access Provider. The Access Provider must state permission for the applicant to:

- enter the Commonwealth area
- take samples from the biological resources of the area
- remove samples from the area

Such written permission has effect only if a permit for the proposed access is issued by the Minister for Sustainability, Environment, Water, Population and Communities.

The permit application must include a copy of a Statutory Declaration that is to be provided to each Access Provider. A statutory declaration is used to ensure the researcher makes a legally binding commitment that they are accessing biological resources solely for non-commercial purposes and restricts them from transferring those resources to others. This statement carries the weight of the *Statutory Declarations Act 1959* (Commonwealth), for which it is an offence to make a false declaration. This has been used as a compliance safeguard for the process.

Access Providers are specified by the EPBC Regulations. In most cases, this is likely to be the Commonwealth, represented by the Protected Area Policy and Biodiscovery Section of DoE.

Permit applications

Permit applications must be made by completing an application form, including information such as:

- **Personal details**: Full name, business address, postal address, and contact details of each person to whom the permit is to be issued
- **Access provider**: Name of each Access Provider. If an Access Provider is the Commonwealth or a Commonwealth agency, the name of the Commonwealth department or agency that administers the Commonwealth area in which the access is proposed
- **Other people involved**: Details of any other person for whose benefit access is sought or who proposes to use the samples obtained
- **Qualifications**: Details of the relevant qualifications or experience of each person proposing to take the action
- **Objectives**: Details of the objectives or purpose of the action, including whether the relevant purpose is commercial/potentially commercial or non-commercial
- **Action**: Description of the action, including the methods to be used to comply with these regulations and to minimise impact on any listed species or native species
- **Species**: Number of listed species or native species that will be affected
- **Resources**: The biological resources to which the applicant seeks access; amount of biological resources that is proposed to be taken, the

proposed use of the biological resources and how access will benefit biodiversity conservation within the area

- **Location and method**: Details of when and where the action will be taken, including the latitude and longitude of the location area; how the access is to be undertaken, including details of vehicles and equipment to be used
- **Involvement of indigenous persons**: Any use that is proposed to be made of indigenous people's knowledge in determining the biological resources to be accessed or the particular areas to be searched, and details of any agreements made with indigenous persons in relation to use of specialised information or information otherwise confidential to the indigenous people of the area
- **Ongoing access**: Whether the applicant thinks that further access to the biological resources will be sought
- **Related applications**: Details of any other application by the applicant for a permit under the regulations.

Source: Adapted from the Commonwealth Department of Environment website: http://www.environment.gov.au/topics/science-and-research/australias-biological-resources/permits accessed 7/3/2014.

In cases where an applicant seeks application for access to biological resources where there is commercial intent, the applicant must establish a benefit-sharing agreement with the access provider. Any benefit-sharing agreement must include as a minimum, the elements described in Box 7.2. This includes a requirement to recognise, value and share benefits if indigenous people's knowledge is to be used. In addition, the applicant has to consult with the owner of land leased by the Commonwealth, before entering into a benefit-sharing agreement with the access provider (in this case, typically the Commonwealth).

There are also requirements for prior informed consent (PIC) where access is sought to indigenous people's land under the regulations: 'If the biological resources to which access is sought are in an area that is indigenous people's land and an access provider for the resources is the owner of the land or a native title holder for the land, the owner or native title holder must give informed consent to a benefit-sharing agreement concerning access to the biological resources' (Article 8A 10(1) of the EPBC Regs).

Box 7.2 Minimum requirements for a benefit-sharing agreement in commonwealth areas

A benefit-sharing agreement must provide for reasonable benefit-sharing arrangements, including protection for, recognition of and valuing of

(Continued)

(Continued)

any indigenous people's knowledge to be used, and must include the following:

(a) full details of the parties to the agreement;
(b) details regarding the time and frequency of entry to the area that has been agreed to be granted;
(c) the resources (including the name of the species, or lowest level of taxon, to which the resources belong, if known) to which access has been agreed to be granted and the quantity of the resources that has been agreed can be collected;
d) the quantity of the resources that has been agreed can be removed from the area;
(e) the purpose of the access, as disclosed to the access provider;
(f) a statement setting out the proposed means of labelling samples;
(g) the agreed disposition of ownership in the samples, including details of any proposed transmission of samples to third parties;
(h) a statement regarding any use of indigenous people's knowledge, including details of the source of the knowledge, such as, for example, whether the knowledge was obtained from scientific or other public documents, from the access provider or from another group of indigenous persons;
(i) a statement regarding benefits to be provided or any agreed commitments given in return for the use of the indigenous people's knowledge;
(j) if any indigenous people's knowledge of the access provider, or other group of indigenous persons, is to be used, a copy of the agreement regarding use of the knowledge (if there is a written document), or the terms of any oral agreement, regarding the use of the knowledge;
(k) the details of any proposals of the applicant to benefit biodiversity conservation in the area if access is granted; and
(l) details of the benefits that the access provider will receive for having granted access.

Source: Article 8A 08 of the EPBC Regs, accessed 1/3/2014.

Since 2006 when the regulations came into force, there have been dozens of non-commercial permits issued and only a handful of commercial permits. The only organisation to have received commercial permits is the Australian Institute of Marine Science (AIMS). The number of permits issued has generally increased and includes the following:

- 1 non-commercial permit in 2006;
- 9 non-commercial permits in 2007;
- 27 non-commercial permits in 2008;
- 16 non-commercial permits in 2009;

- 27 non-commercial permits and 1 commercial permit in 2010;
- 33 non-commercial permits in 2011;
- 45 non-commercial permits and 2 commercial permits in 2012; and
- 46 non-commercial permits in 2013.

The growing numbers suggest that the permit system for non-commercial permits is not onerous. The low number of commercial permits is probably due to at least 2 factors: Commonwealth jurisdiction areas are relatively small and so researchers could easily be choosing to access samples from other locations in the Australian states and territories that do not explicitly require a permit; or this may simply be due to the shift in biodiscovery trends (particularly in pharmaceuticals) whereby natural products discovery is often being undertaken by universities and in non-commercial contexts, but with the potential to become commercial if something interesting is found.

The AIMS is an interesting case. For more than a decade AIMS has been sampling shallow water biodiversity from Australasia and has built up several important reference collections. For its research, which focuses on tropical marine waters, it has the following collections:

- The world's largest collection of coral cores, collected from the East coast of Australia for research on past climatic conditions;
- The AIMS bioresources library which contains about 20,000 entities, including extracts from over 7,600 samples of marine micro-organisms, and over 9,000 cryopreserved marine derived micro-organisms. AIMS is in the process of developing a sophisticated system to make the library more accessible for national screening networks of researchers conducting biodiscovery research;
- Marine sediment samples from more than 2,500 locations across northern and eastern Australia, which have been analysed for chemical makeup (AIMS, 2014).

Under the Commonwealth ABS regulations, it specifies that the Minister may exempt collections from the regulations if those collections can satisfy them that they are CBD compliant (much like the European Union is proposing to do for Registered Collections that would be recognised as CBD and Nagoya Protocol compliant). Two entities: the Australian National Botanic Gardens and AIMS have been deemed by the Minister to be a 'trusted collection' and so may issue PIC and MAT for access to biological resources in their collections (Burton and Evans-Illidge, 2014). These authors note the recent success of this approach in relation to R&D on 'Chondropsins' as a therapeutic anti-cancer lead (from an Australian sponge collected and held at AIMS). The genetic resources were supplied by the bioresources library at AIMS to the US National Cancer Institute and will continue to be supplied by AIMS in a legally compliant and sustainable supply as the research progresses (Burton and Evans-Illidge, 2014).

Access to biological resources in the Northern Territory

The Northern Territory *Biological Resources Act (2006)* (NT BRA) is quite similar to the EPBC Act and regulations, but uses slightly different terminology and has some different terms. The NT law defines bioprospecting as 'the taking of samples of biological resources, existing *in situ* or maintained in an *ex situ* collection of such resources, for research in relation to any genetic resources, or biochemical compounds, comprising or contained in the biological resources' (NT BRA, S5(1)). The law excludes the taking of certain biological resources, for example, human biological tissues, and also biological resources collected for purely commercial or 'use' purposes (e.g. collecting firewood or peat, wild-flowers, recreational or commercial fishing and others).

In the Northern Territory, the researcher must apply for a permit from the CEO of the appropriate authority (currently the Research and Innovation section of the Department of Business, but also in other cases it may be the Agency responsible for issuing permits under the *Territory Parks and Wildlife Conservation Act* or the Agency responsible for issuing permits under the *Fisheries Act*). Part of the permitting process includes a determination of who is the resource access provider. If it is the 'Territory', then the CEO is charged with negotiating a benefit-sharing agreement (in consultation with other relevant agencies). If the research access provider is a different type of land-holder, there are require-ments for negotiation and PIC of a benefit-sharing with those land-holders (see Box 7.3). There is no separation of non-commercial and commercial research under this Act, as opposed to the Commonwealth approach.

Box 7.3 Resource access providers under the NT Biological Resources Act (2006)

(1)   Resource access provider, for biological resources in the Territory to which this Act applies, means the following:

    (a)   for freehold land – the owner of the fee simple (including where the land is subject to a lesser interest such as a lease or licence);

    (b)   for Aboriginal land – the owner of the fee simple (the Aboriginal Land Trust established under the *Aboriginal Land Rights (Northern Territory) Act 1976* (Cth);

    (c)   for an Aboriginal community living area – the owner of the fee simple (an association within the meaning of the *Associations Act* or an Aboriginal association within the meaning of the *Aboriginal Councils and Associations Act 1976* (Cth));

    (d)   for land subject to Native Title (exclusive possession) – the registered native title body corporate;

    (e)   for land held under Park freehold title – the owner of the fee simple (the relevant Park Land Trust established under the *Parks and Reserves (Framework for the Future) Act*);

(f) for Crown land (including land subject to a Crown term lease or Crown perpetual lease) – the Territory;

(g) for land subject to a lease under the *Special Purposes Lease Act* – the Territory;

(h) for land subject to a pastoral lease under the *Pastoral Land Act* – the Territory;

(i) for Territory waters – the Territory.

(2) A bioprospector must make any necessary arrangements for physical access to the resource with the person who controls the physical access.

Source: NT Biological Resources Act (2006) Section 6.

The terms of a benefit-sharing agreement are quite broad, but require 'reasonable benefit-sharing arrangements, including protection for, recognition of and valuing of any indigenous people's knowledge to be used' (NT BRA S29). The law also requires specification of relevant details about the resources to be extracted, the quantities, if traditional knowledge is to be used, details of any potential third-party transfers and details of what benefits are to be shared. There are also interesting provisions for establishment of retrospective benefit-sharing agreements where biological resources are originally taken for other purposes, but are then researched for biodiscovery purposes (S.30). The CEO is also required to issue certificates of provenance relating to samples of biological resources, which are collected from information held in a register from each permit application. The law specifies penalties for people who bioprospect in the Territory without a permit, however, it does not yet include 'user measures' to ensure that local researchers in the Territory are in compliance with foreign bioprospecting regulations (as required under the Nagoya Protocol once it comes into force). It is likely that these user measures will be adopted in Australia gradually across the states and territories as it prepares to ratify the Nagoya Protocol. Box 7.4 provides one example from the Northern Territory that was highlighted as a successful agreement by a past administrator of the NT BRA.

Box 7.4 A bioprospecting agreement between Proteomics International and the NT Government

The Western Australian drug discovery company Proteomics International spent several years trying to establish a commercial bioprospecting agreement before being contacted by the NT Government. Under the NT Biological Resources Act, an agreement has been established to allow the company to hire experts to collect a number of arthropods – spiders, scorpions and

(Continued)

(Continued)

centipedes – and investigate their venom for bioactive compounds that might be used in human therapeutics. They have chosen to investigate venoms because of the evolutionary set of peptides and proteins that the arthropods have developed for defence or attack. The company is seeking to identify a range of bioactive peptides, which could include neurotoxins, analgesics, anti-microbials and others.

The bioprospecting arrangement appears to be a mutually beneficial one. The agreement was made with the NT Government (represented by the Department of Business, Economics and Regional Development – now just Department of Business) for a defined list of fauna. A permit was then issued after the negotiation of a benefit-sharing arrangement. The collection activities occur in Central Australia near Alice Springs where a professional collector identifies appropriate individual arthropods to sample. Under the arrangement, no endangered species are collected, and only a limited number of samples are made. The samples are then transferred to Proteomics International where the venoms are screened for peptides. Under the agreement, if new biological entities are discovered and commercialised, there is a single digit royalty payment back to the NT Government and data sharing of the research results. The data shared is for taxonomy and phylogentic studies, which may be of value to entomologists and for future biological surveys. In addition, specimens are deposited with the Western Australian State Museum for formal identification and archiving where appropriate.

Sources: Evans, N. (24/6/2008) *'NT Signs Off on First Biodiscovery Agreement'*. Available at: *BiotechnologyNews.net,* accessed 14/6/2011; Epstein, B. (14/4/2009) 'A new search for natural peptides yields up to 5 times more potential drug leads'. *Media Release – Proteomics International.* Available at www.proteomics.com.au, accessed 14/6/2011, Lipscombe, R. Personal Communication, 24/6/2011.

The Northern Territory Government was in the process of consulting with stakeholders and reviewing the BRA in 2011 and 2012. Since that time, there has not been any provision of information on the permitting and ABS process, nor is there any detail about it on the NT Department of Business website at the time of writing. This raises the concern that even if researchers and companies were intending to comply with these permits and requirements, it is very difficult to find information about them.

Apart from aspects of the Commonwealth and NT regulations, there is a distinct possibility of issues arising from 'change of intent' regarding biological resources that are publically available on a commercial basis. If a researcher accesses biological resources for commercial purposes (use, sale, trade) and then changes their intent towards R&D, particularly commercially oriented R&D, then it may be difficult for the Australian authorities to pursue these researchers. Under the NT Act this might be undertaken via the retrospective bioprospecting provisions (although in practice it is not clear how the authorities would enforce this). Under the EPBC Act and regulations, there are penalties for failure to obtain a permit, and also penalties for making a false declaration if research of a commercial nature is undertaken with only a non-commercial permit.

The 'Kakadu plum' case highlights this point. The cosmetics company Mary Kay Inc. had applied for WIPO PCT patent application number WO/2007/084998 on 'Compositions comprising Kakadu plum extract or açaí berry extract' on 19 January 2007, for a skin care cream. Being a WIPO filed international patent, it had national filings including: Australian patent application number 2007205838, US patent application number 11/624985, European Patent Office application number EP20070710236 and others (Robinson, 2010a). The patent application sparked interest in Australia among the media and also in parliament and the departments administering ABS regulations. While the patent was being examined the author filed a section 27 submission to IP Australia regarding the novelty and obviousness of the patent application. To contextualise, the Kakadu plum (*Terminalia ferdinandiana*) has been used as a food for hundreds and possibly thousands of years by Indigenous Australians in the Northern regions of Australia. It is one of the world's highest known sources of Vitamin C or ascorbic acid – a powerful antioxidant (along with Açai berry, found endemically in Brazil). A number of Aboriginal corporations and organisations noted at the time in media statements that the plant has some cultural significance – it is mentioned in dreaming stories – and that it was also used by a number of smaller companies that employed indigenous people (although producing extracts or powder from Kakadu plum for use as a food or food additive). It has also been well documented by Gorman et al. (2006), Woods (1995) and Brand et al. (1982) that the Kakadu plum has a history of traditional use as both a food and medicine by Indigenous Australians. These details are also noted in US patent number 7,175,862 on a method of preparing Kakadu plum powder (Robinson, 2010a). In addition, members of the public flagged that companies such as 'Red Earth' which produced skin-care products in the past have also used Kakadu plum extract as an active ingredient in their skin care creams. The examination subsequently reported evidence of prior art and obviousness to those trained in the art (e.g. the field of dermatology) and the filing company ultimately withdrew the patent application in Australia.

Upon further analysis, there was some concern in this case if a permit has been obtained and a benefit-sharing agreement established. The application for a patent is a de facto claim that R&D has been made, since patents require a degree of innovation to be granted. Given the nature of Australia's ABS laws, regarding benefit-sharing for access to 'biological resources' where there is commercial intent and where the resources have been collected from certain areas, the allocation of a permit to Mary Kay Inc. seems pertinent. The natural distribution of the Kakadu plum is largely across the top of the Northern Territory and northern parts of Western Australia, as well as potentially in isolated parts of northern Queensland (Cunningham et al., 2009). There are also a few suppliers of commercially harvested (wild harvest collected) Kakadu plum in the different states and territories in Australia, for personal or commercial use. Initially it was unclear if Mary Kay Inc. had obtained samples from the field or from a supplier, and contacts in the most relevant state and territory governments also indicated that they had not specifically issued a permit to Mary Kay Inc. What seems most likely to have happened is that the company may have obtained supplies from a commercial supplier and conducted some R&D on those samples. In an interview with a radio news reporter – Claire Atkinson of the Special Broadcast Service (SBS) – which went to air on 30 March 2011, a representative of Mary Kay Inc. told the interviewer that the company uses 'an ethical Australian supplier to harvest the Kakadu plum in the Northern Territory' and that 'the harvesting takes place annually near the Kakadu National Park under a licence issued by the Australian Government' (Atkinson, 2011). The key questions here are about 'supply for what intent' and 'when did supply for R&D occur?' While this supplier may be licenced for certain commercial activities, based on discussions with government officials administering ABS in the Commonwealth and Northern Territory Governments it seems likely that they are not permitted to conduct R&D on Kakadu plum – these contacts could not specifically identify a permit that 'fit this description'. However, under these laws, and given that Mary Kay Inc. is not physically located in Australia (its headquarters are in the United States), it may not be possible to require the company to come into compliance with the ABS laws or to penalise them if indeed there has been a breach of an Act. Rather it may be the supplier that has breached their permit conditions, given that permits usually specify what they are and are not allowed to do with the materials. Otherwise, the supplier may be sourcing product from their own property, in which case the researchers should have obtained a permit from the NT government and established a benefit-sharing agreement with that access provider. There is also the possibility that the company initially acquired the biological resources for R&D from overseas, since there has also been speculation of plantations being established in Brazil by another company (Robinson, 2010b). In addition, Mary Kay Inc. may have obtained the original materials for R&D purposes before the entry into force of these laws. These series of questions and issues in this case demonstrates the considerable potential for confusion about and either unintentional or deliberate avoidance of access permits and benefit-sharing under ABS laws. The Nagoya Protocol will seek to address these issues, for example, through the clearing house mechanism which will require certificates of compliance or permits to be registered with the clearing house. However, this will not apply retrospectively and is still not in force at the time of writing. It will also not necessarily receive information from non-parties (e.g. the United States is only a signatory to the CBD and seems unlikely to ratify the CBD and Nagoya Protocol in the foreseeable future), nor is it likely to capture information from trade from ex situ collections, especially private and commercial collections, unless or until those countries accede to the Nagoya Protocol and incorporate requirements for this reporting under their national laws.

Other states and examples

Queensland developed a *Biodiscovery Act 2004* which has some differences to the Commonwealth and Northern Territory laws. While the law establishes a regime facilitating ABS with regard to biological and genetic resources, it has been criticised for its lack of reference to traditional knowledge of indigenous communities or consideration of indigenous communities as access providers. As Collings and Evans (2009) noted, the law does not explicitly require benefit-sharing as of right with Aboriginal and Torres Strait Islander people, does not provide any recognition for their traditional knowledge and does not require PIC or MAT with relevant traditional owners and traditional knowledge-holders. These authors noted that 'while this has not prevented the adoption of ABS agreements between Indigenous groups and pharmaceutical or other resource-accessing parties, there is no requirement that an agreement be drawn up where private entities utilise traditional knowledge' (Collings and Evans, 2009). This lack of consistency between the states and territories may lead to the potential for 'access shopping' between the states to find the lowest regulatory bar for legal sampling or access. Indeed, researchers may well choose to go to the southern states in Australia which have been slow to develop ABS policies, procedures or regulations, and where they can probably obtain samples of biological resources in the wild or from private collectors, nurseries or universities without the same regulatory hurdles. Indeed, there is probably nothing currently in place to stop someone from obtaining indigenous or traditional knowledge about a plant like Kakadu plum from the public domain or from indigenous communities in the northern states, but then seeking to obtain the biological resource from a supplier in New South Wales. At present there is a call for submissions by IP Australia regarding exactly this issue, seeking public and expert opinions on ways to better protect indigenous or traditional knowledge either using existing IP tools or through the creation of new tools.

Despite the regulatory issues, Collings and Evans (2009) do, however, noted the establishment of a relatively positive partnership between researchers at Griffith University in Queensland and the Jarlmadangah Buru people of the Kimberley region in Western Australia (which developed prior to the establishment of the law). The benefit-sharing agreement is related to indigenous knowledge of the bark of the *Barringtonia acutangula* mangrove plant, traditionally used to hunt and trap fish for food. In the early 1990s, Prof Ron Quinn of Griffiths University learned of the plant's analgesic potential after hearing about an Aboriginal man in the Kimberley region of north-western Australia whose finger had been bitten off by a crocodile – the man took the bark of the tree, chewed it around in his mouth and then put it on the wound (Skatssoon, 2004). Over a number of years, following the formation of a partnership with the Jarlmadangah Buru, researchers from Griffith University developed an analgesic compound from the bark of the plant (Collings and Evans, 2009). The agreement between Griffith University and the Jarlmadangah Buru provides for returns from any commercial development to be split 50:50 between the Aboriginal group and Griffith University (Laird et al., 2008; Skatssoon, 2004). This is a very high split of return (especially if return of revenue and not profit) for an ABS agreement, given that royalties are often specified as a single digit amount (of profits). This high level of return arguably provides a strong recognition of the contribution made by the traditional knowledge to the isolation of specific compounds and development of a useful drug.

Another project discussed at length by Laird et al. (2008) is the Griffith University Astra Zeneca partnership. Griffith's partnership with Astra Zeneca was launched in 1993, renewed in 1998 and again in 2002, and was concluded in 2007. Over this time, Astra Zeneca invested more than AUD$100 million in the program (Laird et al., 2008, p. 20). The partnership allowed the collection of an extensive array of samples by the Queensland Herbarium from Queensland, but also other overseas and interstate locations, for further testing by high throughput screening. As of late 2012 at the Oceania Biodiscovery Forum held at Griffith, Prof Quinn of the Eskitis Institute indicated that despite many leads, there has been no commercial product achieved to date. Laird et al. (2008) noted the many benefits of the partnership, particularly through the large commercial investment in the building of infrastructure at Griffith, and the funding of maintenance and sampling undertaken in the project. However, they also noted that there were a range of media-based public and indigenous concerns expressed including:

- The exclusivity of the agreement with Astra Zeneca;
- The 'locking up' of Australian resources by multinational companies through such agreements;
- The potential use of traditional knowledge as a lead towards new drugs without acknowledgement;
- The potential collection from aboriginal lands or lands soon to be subject to Native Title;
- The fairness of royalties to be paid back to Australia in the event of a commercial product (Laird et al., 2008, pp. 26–27).

A final case that is worth mentioning, not so much as an 'ABS' agreement, but rather as a successful biodiscovery activity is the research that has been conducted by Prof Jim Aylward who was at the University of Queensland. He examined the use of Milk Weed or Radium Weed (*Euphorbia peplus*), which is a commonly found weed (not indigenous to Australia) which has been used for possibly centuries as a home remedy to treat sunspots (Gaffney and Wood, 2009; Burton, pers. comm. 2012). Researches on a gel made from extracts of the sap of the plant have led to further development of a drug for the treatment of certain types of skin cancer. Three phases of clinical trials have reportedly been passed in Australia and a number of trials also passed in the United States. This has then led to the acquisition of 'Peplin Inc' – the company that Prof Aylward established to undertake the research in 1998, with the support of the Queensland Institute of Medical Research. In 2009, the Danish company LEO

Pharma (2009) announced its acquisition of Peplin for $287.5 million in cash. The case highlights that natural products discovery is still relevant to the pipeline of pharmaceutical drug development. It also highlights the potential revenue that can be obtained by researchers and institutions who do this sort of research, and thus it follows that benefit-sharing might also extend in certain circumstances. In this particular case, it is unclear if there should be additional beneficiaries – if the genetic resource is a widely spread weed and there is common knowledge about its potential use, can the researchers and companies involved reasonably be expected to share benefits? Would the original 'access/resource provider' have any claim in this case with or without a regulatory system for ABS in place? The answers to these questions are unclear in this and a number of other similar cases.

References

Atkinson, C. (2011) 'Anger over Kakadu plum patent application' SBS Radio World News Australia – on air 30 March 2011.

Australian Government Department of Environment and Heritage (now DoE). (2002) 'Nationally Consistent Approach for Access to and the Utilisation of Australia's Native Genetic and Biochemical Resources'. Canberra. Available at: http://www.environment.gov.au/resource/nationally-consistent-approach-access-and-utilisation-australias-native-genetic-and, accessed 14/7/2014.

Australian Government Department of Environment (DoE). (2014) *'Access to Biological Resources in Commonwealth Areas'*. Available at: www.environment.gov.au/topics/science-and-research/, accessed 5/3/2014.

Australian Government Department of Foreign Affairs and Trade (DFAT). (2008) *'Australia's Environment at a Glance'*. Available at: http://www.dfat.gov.au/facts/env_glance.html, accessed 14/7/2014.

Australian Institute of Marine Science. (2014) ' *"Facilities" and "Collections" Webpages'*. Available at: http://www.aims.gov.au/docs/about/facilities/facilities.html and http://www.aims.gov.au/docs/research/biodiversity-ecology/collections/collections.html, accessed 18/3/2014.

Brand, J.C., Rae, C., McDonnell, J., Lee, A., Cherikoff, V. and Truswell, A.S. (1982) 'The nutritional composition of Australian Aboriginal bush foods'. *Food Technology in Australia,* 35, 293–298.

Burton, G. and Evans-Illidge, L. (Forth. 2014) 'Emerging R&D law: The Nagoya protocol and its implications for researchers'. *ACS Chemical Biology,* issue forthcoming.

Burton, G. pers. comm. (2012) *Comments on the LEO Pharma acquisition of Peplin at the Oceania Biodiscovery Forum.* Brisbane.

Collings, N. and Evans, H. (2009) 'Access and benefit-sharing – protecting biodiversity and indigenous knowledge'. *Indigenous Law Bulletin,* 7(14), 11.

Cunningham, A.B., Garnett, S., Gorman, J., Courtenay, K. and Boehme, D. (2009) 'Eco-enterprises and *Terminalia ferdinandiana*: "Best Laid Plans" and Australian Policy lessons'. *Economic Botany,* 63(1), 16–28.

Gaffney, A. and Wood, K. (2009) 'Milk weed and skin cancer. On mornings with Annie Gaffney'. *ABC Radio Queensland.* Available at: http://blogs.abc.net.au/queensland/2009/05/milkweed-and-sk.html, accessed 18/3/2014.

Gorman, J.T., Griffiths, A.D. and Whitehead, P.J. (2006) 'An analysis of the use of plant products for commerce in remote Aboriginal communities of Northern Australia'. *Economic Botany,* 60(4), 362–373.

Laird, S., Monagle, C. and Johnston, S. (2008) *Queensland Biodiscovery Collaboration: The Griffith University AstraZeneca Partnership for Natural Product Discovery. An Access & Benefit Sharing Case Study*. UNU-IAS, Yokohama.

LEO Pharma. (2009) *'LEO Pharma to Acquire Peplin for $US287.5m'*. Available at: press release from www.leo-pharma.com/home/LEO-Pharma/Media-Centre/News/News-2009/2009-sep-03-LEO-Pharma-to-Acquire-Peplin-for-US$287.5m.aspx, accessed, 18/3/2014.

Robinson, D. (2010a) *Confronting Biopiracy: Cases, Challenges and International Debates*. Routledge/Earthscan, London.

Robinson, D. (2010b) 'Traditional Knowledge and Biological Product Derivative Patents: Benefit-Sharing and Patent Issues Relating to Camu Camu, Kakadu Plum and Açaí Plant Extracts – a discussion paper'. *United Nations University Traditional Knowledge Inititative (UNU-TKI)*. Available at: http://www.unutki.org/news.php?doc_id=174, accessed 09/09/2014.

Skatssoon, J. (2004, 20 June) 'New in science – Mangrove bark dulls the pain'. *Australian Broadcasting Corporation, Science Online*. Available at: http://www.abc.net.au/science/articles/2004/06/23/1138407.htm, accessed 19/3/2014.

Woods, B. (1995) A study of the intra-specific variations and commercial potential of *Terminalia ferdinandiana (Excel)* (the Kakadu Plum) (M.Sc. Thesis). Northern Territory University, Darwin.

8 The Santo 2006 Global Biodiversity Expedition, Vanuatu

Background

Santo 2006 was a scientific expedition to document the flora and fauna, both marine and terrestrial on the Island of Espiritu Santo in Vanuatu. The expedition involved a natural history inventory, and also a taxonomic inventory (including alien species), intended to provide a baseline of biodiversity for scientists to monitor on the island due to impacts such as climate change (Nari, interview, 12/6/12). The expedition involved collaboration between the French National Museum of Natural History (MNHN), the French Institute for Research for Development (IRD), Pro-Natura International (an NGO which studies tropical forest canopies) and involved the Ministry of Lands of the Government of Vanuatu as the umbrella organisation in Vanuatu. Over 100 participants from 15 countries were involved in fieldwork focused primarily between August and December 2006 (Bouchet et al., 2006). As described by some of the project leaders:

> The land area of Santo and its marine fringes host a mosaic of habitats that have remained largely unexplored. Santo's complex ecological diversity and its geographical position within the archipelago[sic] of Melanesia suggest a very high level of biological diversity. Much of its flora and fauna are still to be discovered, most notably in mega-diverse groups like insects and mollus[c]s. Santo lies outside the centers of economic growth, with the consequence that it has been spared the global standardization that is affecting much of the planet.

> Culturally and linguistically, Santo is also uniquely diverse. This biodiversity survey will document all the major environments (offshore deep-sea, reefs, caves, freshwater bodies, mountains, forest canopies) and will also address issues on how indigenous biodiversity has been impacted by 2,500 years of human presence.

> (Bouchet et al., 2006, p. 2)

Thus the project sought to document previously unknown taxa, to relatively systematically describe the full range of ecosystems on the island, and to understand human impacts as well as potential conservation challenges. Therefore the project was purely scientific and academic, with no commercial partners or intent from the outset.

Access and R&D conducted

The expedition was primarily a taxonomic inventory of species on the island of Santo, intended to answer academic and practical environmental change questions. Bouchet et al. (2008, p. 404) explained that it was the 'gap between taxonomy and conservation that the SANTO expedition has attempted to bridge'. The scientific program consisted of 4 major themes centred on their major sampling facilities and intended to investigate all habitats of the island (deep offshore, coral reefs, continental and marine caves, lowland and highland forests and rivers). The Marine Biodiversity theme was centred around Luganville in the Southeast of the Island and near the Maritime college which acted as a base for the research theme. The Forests, Mountains, Rivers theme was located in the surrounds of Butmas village (central Santo), Penaoroa (a relatively inaccessible area in the Northwest of Santo) and near Kerepoa (near the central West coast of Santo). The Karst Caves theme was based near Luganville, and research was conducted at multiple locations along the east coast of Santo in particular, as well as some locations in the far Northwest and Southwest of the Island. This theme included surveys of underwater inland and coastal caves. The Fallows and Aliens theme focused on the Southeast part of the Island near Luganville, which is the most developed and disturbed area of the island. It was also conducted at Vatthe Conservation Area in the north of Santo on the inland coast of Big Bay, where historical plantations and agriculture have affected the modern day ecology of the area. The project also notes a 'Cultural Perceptions of Biodiversity' theme from the outset; however, this appears to have been scaled back due to some underlying concerns in Vanuatu (discussed below).

The main questions posed by the project leaders included: what is the real magnitude of biodiversity when the richest habitats and most diverse taxa are considered? What is the share of rare species and species assemblages? What is the spatial distribution of biodiversity? and how do we evaluate site representativeness at an 'ecoregional scale' (Bouchet et al., 2006)? The project involved sampling of species, but this sampling was not apparently conducted for commercial purposes. The project leaders noted that during negotiations and consultations with government agencies in Vanuatu prior to the expedition, 'reservations arose on some of the objectives of the expedition in the field of ethnobotany – especially ethnopharmacology. Given that this was a very minor part of our project ... and to avoid being suspected of biopiracy, we decided to remove this component' (Bouchet et al., 2011). The agreement terms respond to some of these concerns as further discussed below.

The expedition team and a number of authorities in Vanuatu reached a Memorandum of Understanding in November 2005. This was circulated to various government departments and the Luganville Provincial Secretary for comment (Path, interview, 13/6/12). The agreement was ultimately signed on 24 March 2006 by the then Minister of Lands, Maxime Carlot, and the then director of the Paris Museum, Bertrand-Pierre Galey, who represented the Santo 2006 expedition team. This agreement then provided the underpinning for a research permit issued collectively to all members of the expedition, signed on 2 June 2006 by Ernest Bani, director of Vanuatu's Department of Environmental Protection (Bouchet et al., 2011). The agreement signed between the Government and MNHN asserts that the expedition 'commits to collect information and specimens for academic and management purposes only' and committed to the Vanuatu Cultural Centre giving its prior informed consent (PIC) to any publication that might contain elements of indigenous or traditional knowledge gathered by the project (Bouchet et al., 2011, p. 545).

The agreement also commits to the following terms, among others:

- 'taking steps to prevent the use of this information and specimen for commercial purposes after the end of the period of the present agreement'
- The housing biological samples collected during the expedition at the Forest Department (plants), Fisheries (fish), Cultural Centre (fossil vertebrates), and insects and other biota of quarantine interest at the Quarantine Department.
- The assertion of intellectual property rights of the Government of Vanuatu over any data created from the expedition, except for indigenous or traditional knowledge which belongs to the person providing it. The parties to the agreement are free to use the data for academic research.
- The agreement was signed for effect over a period of 3 years from March 2006 (MNHN – Vanuatu Agreement, 2006).

After government permits were issued and agreements made in Port Vila, the expedition team also sought permissions from leaders on the island of Santo. As described by the project leaders:

> In Santo, as in the rest of Vanuatu, the real power to accept visitors comes from the communities themselves, at the level of customary chiefs and villages. We therefore had to contact all members of these communities and inform them about our projects with the support of the traditional chiefs, as well as the field workers of the Vanuatu Kaljarol Senta (Cultural Centre – VKS) and the members of parliament for West Santo, Sela Molissa, who personally took part in informing the people of "his" constituency.
>
> (Bouchet et al., 2011, p. 532)

A large meeting was held in Luganville with the chiefs of the province with many people coming to hear about the objectives of the expedition in English and Bislama. Specific activities in villages required PIC from village leaders (Nari, interview, 12/6/12; Path, interview, 13/6/12; MNHN – Vanuatu Agreement, 2006). Rufino Pineda, the national coordinator for the expeditions, indicated that he had digital copies of the PIC agreements signed by local chiefs but could not locate them at the time of interview (Pineda, 12/6/12). Interviews in Butmas village and Matantas/Vatthe village confirmed that local permissions were obtained (Serei Maliu, interview, 14/6/12; Chief Solomon, interview 14/6/12; Purity Solomon, interview, 14/6/12; Bill Tavine, interview, 14/6/12).

Benefit-sharing

From the outset, it was made clear that the project had non-commercial scientific aims. Therefore the primary benefits directed towards Vanuatu were indirect monetary benefits, or non-monetary benefits, including the exchange of scientific information, provision of training and similar. The agreement specified that:

> The Museum [MNHN] commits to involve Vanuatu Cultural Centre Fieldworkers and community workers for data collection, as well as, to the largest extent possible, ni-Vanuatu biodiversity officers, technicians, and students. It commits to do its best to continue after the field season the training of ni-Vanuatu biodiversity officers, technicians, and students, by facilitating their access to grants and training in institutions in France and New Caledonia
>
> (MNHN – Vanuatu Agreement, 2006, p. 2)

We interviewed several people in Butmas and Matantas villages. These are 2 of the villages in the centre of Santo and the North, respectively, from which researchers based themselves and went out into surrounding areas for their surveying. From the interviews, respondents indicated that there was not much training of local people and guides about the species being collected. Rather the researchers came, stayed for a short time and then left (Serei Maliu, interview, 14/6/12; Chief Solomon, interview 14/6/12; Purity Solomon, interview, 14/6/12; Bill Tavine, interview, 14/6/12). An informant involved in the project noted that it was difficult to find local people with basic relevant training who were interested in being involved in the research, hence the low level of training. In Matantas village, the community has received a copy of the Natural History of Santo book and some illustrative posters of local biodiversity for their school as supplied by Mr Rufino Pineda (Chief Solomon, interview 14/6/12; Purity Solomon, interview, 14/6/12; Bill Tavine, interview, 14/6/12). Butmas village, a remote village in the mountains, was a research site that was occupied for only approximately 1 week by the researchers (Figure 8.1). The village Chief indicated he had not received the posters and post-expedition information yet (Serei Maliu, interview, 14/6/12). At the time of interview, Mr Pineda indicated that he would endeavour to ensure that additional posters were distributed to relevant villages that may have missed out (Pineda, 12/6/12).

It was also agreed that biological sample duplicates (where they were collected–sometimes only individual specimens could be found) would be

Figure 8.1 One of the research camps in 2006 was based at Butmas Village, Santo, Vanuatu

Source: Robinson, 14/6/2012.

housed in Vanuatu at the Cultural Centre in a wing focusing on natural history that is yet to be built, and at other relevant departments including Fisheries and Forestry. To date, many of these samples are still housed in foreign facilities, including the Museum in Paris, because the facilities in Vanuatu are not yet established and taxonomic work is still being undertaken (Nari, interview, 12/6/12). Duplicate samples were also sent to the University of South Pacific and a US-based institute (Pineda, interview, 12/6/12; Mr Pineda indicated that he believed that it was the Smithsonian).

The project team has not yet passed on scientific literature based on the expedition to the Vanuatu Cultural Centre/National Library or the Department of Environmental Protection, as was originally agreed (Kalfatak, interview, 11/6/12; Hickey, pers. comm. 11/6/12; Norman, pers. comm. 11/6/12). The agreement specifies a website to be established to allow free access to articles produced as an outcome of the research. A website containing many freely accessible photos and limited explanatory documents related to the expedition is still available (www.santo2006.org). The French IRD website still contains some content on the research activities. Bibliographies are provided but not the actual scientific publications.

Russell Nari, formerly the Director-General of the Department of Lands (interview, 12/6/12), noted that many copies (approximately 400) of the book have been supplied to the Vanuatu Government, schools and stakeholders and some scientific papers, but not many yet. He noted that these were recently published and there are still many publications being produced (see Table 8.1 for a summary of benefits).

Although it cannot really be considered 'benefit-sharing', basic payment of local communities, guides and assistants occurred during the expedition. This represents a temporary source of income that might not have otherwise existed (Serei Maliu, interview, 14/6/12; Chief Solomon, interview 14/6/12; Purity Solomon, interview, 14/6/12; Bill Tavine, interview, 14/6/12). There were sometimes issues over the amounts paid or the work expected (Kalfatak, 11/6/12). One of the participants anonymously complained about the low wage paid for their research assistant work (anonymous, interview, 12/6/12). In the Vatthe conservation area, researchers paid 600 Vatu (approximately $6) to be used for community conservation as a one-off fee for entry. Purity Solomon from Matantas village noted that this was the only payment or contributions made to the community – other payments were made to individuals for their work as porters or field assistants (interview, 14/6/12).

Table 8.1 Summary of monetary and non-monetary benefits related to the Santo expedition

Monetary	Non-monetary
Payment of local communities, guides and assistants for accommodation and assistance (not strictly 'benefit-sharing', but rather an indirect benefit).	Inclusion of 10 Ni-Vanuatu participants in the expedition with the scientific team and related training.
Restoration of a boat *Euphrosyne* which was used to transport people from Luganville to the western side of Santo (worth an estimated $125,000), which was until recently the property of the Vanuatu Maritime College and provided transport services to the island.	Production and distribution of approximately 400 copies of the '*Natural History of Santo*' book (in French and English) to authorities in Santo, in Port Vila and to schools.
Direct injection of an estimated 2.4 million Vatu or $26,000 into the economy of Penaoru and villages surrounding it on the west coast of Santo (camp construction, portering, supplies, guides etc). Again this is not strictly 'benefit-sharing' but might be considered incidental benefit.	Contributions to understandings of the local taxonomy of Santo and its surrounding marine areas.
	10 educational posters created on various biodiversity themes, printed as 500 copies, disseminated to every school in the country in the 3 main languages of Vanuatu with EU funding.

Impact

The Expedition claims to have identified 650 species of plants (higher plants, ferns, mosses and liverworts), 350 species of fungi, 1,700 species of terrestrial animals, 1,100 species of decapods crustaceans, 4,000 species of molluscs and 650 species of fish, of which hundreds of new species have been collected and are being described (Waiwo, 2011). Details of the entire expedition have been published as a book, *The Natural History of Santo*, of which many copies have been provided to relevant authorities in Santo and Vanuatu, as well as to schools to 'increase the awareness of schoolchildren on the diversity of life and to introduce them to "science in action"' (Waiwo, 2011). Mr Joel Path, the Secretary General of Sanma Province (including Santo), noted that the book, documentaries and publicity created from the expedition had been good for Santo. He indicated that it had boosted interest in the local environment and had an impact on tourism, particularly ecotourism (interview, 13/6/12).

In addition, EU funding was received for the preparation and publication of 500 copies of posters featuring the biodiversity of the island of Santo to be distributed to schools in Vanuatu. These were both provided to the Ministry of Education for distribution and some were distributed to individual schools by the National Coordinator of the expedition (Pineda, interview, 12/6/12).

While the team was able to include 10 Ni-Vanuatu participants in the expedition with the scientific team, they noted the lack of local academic institutions in the field of biodiversity. Philippe Bouchet, a project leader for the Santo Expedition, noted that:

> Vanuatu is a small country and, overall, we found that the one difficulty in following the spirit and letter of the CBD for the Santo 2006 project was the scarcity of in-country partners. We proactively searched for students, technical officers etc. to take part in the expedition, but were only moderately successful.
>
> (pers. comm. 4/6/12)

Although local field guides were used and paid, there was minimal training of local people, except for some training of the 10 Ni-Vanuatu scientists or participants (generally from government or universities). In terms of training, the project leaders noted that following the expedition, one of the participants (Samson Vilvil-Fare) received a grant from the Territory of New Caledonia to do a Master's degree at the Pierre and Marie Curie University in Paris, in fulfilment of one of the commitments made under the agreement (Bouchet et al., 2011, p. 545).

There was also some investment in infrastructure such as water supplies in the west coast camp site at Penaoru, plus the refurbishment of a boat to be used in the expedition (then given to the Maritime College of Santo) (Pineda, interview, 12/6/12; Path, interview, 13/6/12).

Products of the research and development

The main 'products' of the R&D are scientific papers and *The Natural History of Santo* book. Several papers have been recently published including Gerstmeier and Schmidl (2007), Jaume and Queinnec (2007), Malzacher and Staniczek (2007), Pyle et al. (2008), Golovatch et al. (2008), Ng and Naruse (2007), Kantor et al. (2008), Terryn and Holford (2008) and Bouchet et al. (2008) among others. Most of these have been written as taxonomic descriptions of newly identified species, many of which are named after the island of Santo or the country, Vanuatu. Notably there is little co-authorship of the papers by Ni-Vanuatu people, which the project leaders acknowledge was an unfortunate outcome resulting from limited human capacity in the biological sciences in Vanuatu. Only 2 of the chapters in the *Natural History of Santo* book appear to have been authored or co-authored by Ni-Vanuatu people. There were some frustrations regarding the lack of co-authorship evident in interviews with some of the stakeholders in Vanuatu.

Discussion

The case highlights some of the differences between R&D intended towards commercialisation and research intended towards academic or scientific discovery. When establishing their project, the leaders met some concerns from individuals about whether they had intentions of conducting 'bioprospecting' type activities. As a result, the project was altered so that ethnobotanical aspects were no longer included, and terms were included in the agreement to ensure PIC from the Vanuatu Cultural Centre before any publication related to the traditional knowledge of Vanuatu. Several commitments were made in the agreement with the Government of Vanuatu, including sharing of data and intellectual property, sharing of published documents and photography, provision of duplicate samples to be housed with relevant government agencies and the provision of training to Ni-Vanuatu people. These commitments have been at least partially achieved to date and others are pending (e.g. repatriation of specimens upon completion of the additions to the Cultural Centre, further provision of publications from subsequent examination of collected species overseas, provision of specimens to the Forests Department after the refurbishment of storage cabinets). There appear to have been other indirect benefits to Santo and Vanuatu (although difficult to quantify), including tourist interest in the biodiversity of the island and some of the sites (such as Millennium Cave), as well as the provision of some infrastructure and a refurbished boat to the Maritime College.

The project leaders also seem to have intended to involve greater collaboration from local scientists and technical staff, but due to a lack of human capacity in the biological sciences in Vanuatu, this was limited during the expedition. Instead, the results of the research – largely published taxonomic works – will hopefully provide benefits towards conservation of species and ecosystems in

Vanuatu. In order to successfully utilise this data towards conservation and sustainable use, further training and building of human capacity relating to biodiversity must take place. Some Ni-Vanuatu stakeholders described this as a lost opportunity for further training. For example, Sam Chanel – the Forest Department's chief botanist was hoping that there would be further taxonomic training for Ni-Vanuatu, including for a potential replacement for him when he retires. He noted that it would be useful if the project helped develop a book or papers specifically on forests in Santo, and more basic provisions to facilitate the benefit-sharing. Specifically he noted that some money had been received for refurbishment of cabinets at the forest department, but more was needed and also air conditioning is important to reduce humidity in the room where the cabinets are located (Chanel, 14/6/12).

A few people raised questions about compliance and monitoring of the use of the samples taken into the future. Stakeholders from the Vanuatu Cultural Centre and Department of Environment wondered aloud how they could be sure that third party transfers weren't occurring, or if they did occur that they were only occurring for academic or scientific purposes, not for commercial research. This raises further questions about how dispute settlement might be resolved. The basic Memorandum of Understanding (MoU) between the 2 countries is already expired and so there is no avenue for contract dispute settlement. If a dispute were to arise, it is not clear in the agreement under which jurisdiction this would be settled, or how a decision would be enforced. These are hypothetical questions, but relevant in many of these kinds of international genetic resource transactions. In this case, the agreement at least specifies a number of things – a 'best endeavours' clause that:

> The Museum commits to taking steps to prevent the use of this information and specimen for commercial purposes after the end of the period of the present agreement.

Which provides some assurance of their intent (albeit unenforceable and not monitored by anyone except the NMNH), leaving the future to a situation of trust. The second important element in the agreement is the specification that data collected during the expedition is the:

> . . . intellectual property of the Government of Vanuatu, except in the case of the data that constitutes 'indigenous or traditional knowledge' in which case the data remains the intellectual property of the person(s) who provided it.

These elements of the agreement at least provide a level of assurance of the intent of the researchers – which, given the many taxonomic publications arising still seems genuine – even if the agreement does not hold much legal certainty. In the future, the establishment of 'model terms for MAT', clearer permitting systems and also the CBD Nagoya Protocol Clearing House mechanism should hopefully assist with the provision of greater legal certainty and improved monitoring. At the time of writing, the Government of Vanuatu was in the process of undertaking national consultations and a number of other actions towards ratification of the Nagoya Protocol, having signed in 2011.

References

Bouchet P., Le Guyader H. and Pascal O. (2008) *The Natural History of Santo*. French National Museum of Natural History, Paris.

Bouchet, P., Le Guyader, H. and Pascal, O. (2009) 'The SANTO 2006 Global Biodiversity Survey: An attempt to reconcile the pace of taxonomy and conservation'. *Zoosystema*, 31(3), 401–406.

Bouchet, P., Le Guyader, H. and Pascal, O. (2011) *The Natural History of Santo*. National Museum of Natural History, France, IRD, Pro-Natura, Paris, and Government of Vanuatu.

Bouchet, P., Le Guyader, H., Pascal, O. and Pineda, R. (2006) 'Expedition Santo 2006 Global Biodiversity Survey: From Sea Bottom to Ridge Crests. Project Synopsis.' Available at: http://www.ird.fr/recherche/santo2006/pdf/santo_project_us.pdf, accessed 14/7/2014.

Gerstmeier, R. and Schmidl, J. (2007) '*Omadius santo* sp. nov. from Espiritu Santo, Vanuatu (Coleoptera, Cleridae, Clerinae)'. *Entomologische Zeitschrift*, 117(2), 85–87.

Golovatch, S., Geoffroy, J.-J., Mauries, J.-P. and Vandenspiegel, D. (2008) 'The first, new species of the millipede family Pyrgodesmidae to be recorded in Vanuatu, Melanesia, southwestern Pacific (Diplopoda: Polydesmida)'. *Arthropoda Selecta*, 17 (3–4), 145–151.

Jaume, D. and Queinnec, E. (2007) 'A new species of freshwater isopod (Sphaeromatidea: Sphaeromatidae) from an inland karstic stream on Espiritu Santo Island, Vanuatu, south-western Pacific'. *Zootaxa*, 1653, 41–55.

Kantor, Y., Puillandre, N., Olivera, B. and Bouchet, P. (2008) 'Morphological proxies for taxonomic decision in turrids (Mollusca, Neogastropoda): A test of the value of shell and radula characters using molecular data'. *Zoological Science*, 25, 1156–1170.

Malzacher, P. and Staniczek, A. (2007) '*Caenis vanuatensis*, a new species of mayflies (Ephemeroptera: Caenidae) from Vanuatu.' *Aquatic Insects*, 29(4), 285–295.

MNHN (French National Museum of Natural History) – Government of Vanuatu Agreement for the Santo 2006 Global Biodiversity Survey Expedition. (2006).

Ng, P.K.L. and Naruse, T. (2007) '*Liagore pulchella*, a new species of xanthid crab (Crustacea: Decapoda: Brachyura) from Vanuatu.' *Zootaxa*, 1665, 53–60.

Pyle, R.L., Earle, J.L. and Greene, B.D. (2008) 'Five new species of the damselfish genus *Chromis* (Perciformes: Labroidei: Pomacentridae) from deep coral reefs in the tropical western Pacific'. *Zootaxa*, 1671, 3–31.

Terryn, Y. and Holford, M. (2008) 'The Terebridae of Vanuatu with a revision of the genus *Granuliterebra* Oyama 1961'. *Visaya*, 3(supplement), 1–96.

Waiwo, E. (2011) 'The natural history of Santo book released to Sanma authorities'. *Vanuatu Daily Post*, Available at: http://www.dailypost.vu/content/%E2%80%98-natural-history-santo%E2%80%99-book-released-sanma-authorities, accessed 6/13/2011.

9 The ICBG Papua New Guinea project

Background

As mentioned in Chapter 3, the ICBG projects have been established as a 'unique effort that addresses interdependent issues of biological exploration and discovery, socioeconomic benefits, and biodiversity conservation' (NIH, 2012). In this case the ICBG have provided grant funding for collaboration between the University of Papua New Guinea (UPNG), the University of Utah and University of Minnesota, the Smithsonian Tropical Research Institute (STRI) on the conservation and sustainable use of biodiversity in Papua New Guinea (PNG). Wyeth was originally a partner organisation on the project but has since dropped out (Cragg et al., 2012). The overarching goal of the ICBG project is to improve human health and well-being through a set of programs dedicated to the description, assessment, utilisation and conservation of biodiversity in PNG. This project has been running for approximately 9 years and is in its second cycle of grant funding from the Fogarty International Centre. Interviews and correspondence were obtained for this case study from 3 key informants: Prof. Louis Barrows (University of Utah), Prof. Lohi Matainaho (UPNG) and Dr Eric Kwa (PNG Law Reform Commission, previously involved in the development of the national ABS permit system), as well as relevant literature.

Access and R&D conducted

The biodiscovery activities are broad in scope and are described in the memoranda between the partners as 'a scientific research collaboration to investigate the biological chemical and medicinal properties of the biodiversity of Papua New Guinea and to establish economic value thereof'. The primary activity of drug discovery focuses on HIV and tuberculosis, with source organisms sought from terrestrial endophytic microbes and marine invertebrates. The project seeks to identify new therapeutic medicines from either validated traditional medicines or developed traditional medicines.

The project also has a secondary emphasis on documentation and preservation of traditional medicinal plant knowledge in PNG. Towards this, the

ICBG supports the PNG Ministry of Health's Traditional Medicines Taskforce and student theses to provide pharmacologic validation and chemical standardisation of medicinal plants, and the identification of novel bioactive molecules (University of Utah, 2010). Given the regular loss of traditional knowledge (TK) from cultures around the world, this sort of process is important to retain the collective knowledge. However, it obviously needs to be done in an ethical and culturally sensitive way (discussed below).

A third area of research activity focuses on conservation and biodiversity in a forest dynamics plot established in Wanang, PNG. The plot has been used by the collaborators for analysis of forest dynamics, carbon sequestration, climate effects, botanical surveys and ecosystem studies, including analysis of soil microbes and small plants. This third area particularly has a focus in line with the original intent of the access and benefit-sharing (ABS) provisions of the CBD, towards benefits which assist with the conservation of biological diversity.

Conformance with ABS legislation and permits

In 1998, the PNG Department of Environment and Conservation (DEC) established the PNG BioNET (initially called the PNG Biodiversity Institute): an organisation of PNG scientists and government officials advisory to the DEC on assessment, use and development of PNG biological resources. The ICBG collaboration was influential in bringing together stakeholders in workshops to discuss issues relating to ABS, and the role of the PNG BioNET (Matainaho, pers. comm. 15/12/11; Kwa, pers. comm. 20/11/12).

The Draft PINBio Act seeks to establish PNG BioNET as the national clearing house for all research permits and access to PNG's genetic resources. Despite being a draft law, there is a formal permit procedure in place through PNG BioNET and DEC. The ICBG follows these procedures and appears to go to some additional lengths for local prior informed consent (PIC) as described below. Permit procedures followed by the ICBG include the following:

- A research proposal is submitted to DEC (by both PNG and foreign researchers).
- The proposal is registered with the Wildlife Enforcement Branch of DEC and the PNG BioNET Technical steering committee (made up of several relevant government department representatives) reviews the proposal and endorses or denies it.
- Considerations include that the proposal must be consistent with the CBD and develop an MOU fostering scientific collaboration to build PNG's scientific and research capacity. The MOU should specify joint patents/publications, recognise PNG's sovereignty over natural resources, consider local values and practices.
- Recommendation by PNG BioNET then is submitted to DEC for approval. This is subject to DEC obtaining approval from research

committees at the relevant provincial government level and consultation with local government.

- Approved biological research activities are registered by PNG BioNET Secretariat and DEC who then issue permits.
- Export permits are also required from Agriculture/Quarantine for plants, and from DEC for marine samples.
- Where samples are sought from locally owned land, PIC must be sought from the owners/chiefs. If they approve the research, they should be involved in their activities. This might include payments for local guides and research assistants and basic taxonomic training. Further to this, the ICBG has a standard operating procedure for local PIC and upfront/field benefit-sharing when conducting collecting activities (described below).

Local PIC procedure followed

For obtaining local PIC (and sharing upfront benefits with local people), the ICBG project has a standard operating procedure including the following steps:

- Reconnaissance visits are conducted and permission sought in advance of collection activities. Translators are used where relevant to discuss the potential research and collection activities in local dialects.
- Villagers are hired as research assistants (approximately K10/day or $4.70)[1] and fees paid to stay in the village accommodation (often with families, approximately K35/day or $16.45 per person). Food supplies are donated to feed families and field assistants. Sometimes medicines are also donated.
- An appreciation fee can be paid to the chief, community leaders or land owners of approximately 1,500 per trip. This can be given as cash or (preferably) through the purchase of supplies (e.g. water tanks, building materials).
- A list and number of plants collected are provided to community. An explanation of potential direct and indirect benefits (education, royalty payments to people or conservation trust, global health/conservation understanding) is made.
- Community leaders are informed of risks (e.g. community conflict or discord, cultural erosion through infusion of money, unrealistic expectations of rewards that do not arise, violation of spiritual values of plants or traditional knowledge).
- The team usually makes an offer of running a preventative medicine workshop to the community and offers to conduct an assessment of health needs of the community.
- They offer an economic valuation of their forest area (given a range of use values, potential use values and also carbon sequestration values, amongst others).
- They may inform communities of potential royalty benefits – 25% of those received by UPNG to the community where the 'hit' arises (discussed in more detail below).

- Local villages are provided with conservation outreach material and possible workshops run by partner NGOs.
- When permission is granted a $300 donation is made to the school fund of the village, including support for school costs to financially disadvantaged children of approximately K50-100 per child (or from $23.50 to 47).
- Prompt feedback should be provided to the collection areas, including plant survey data, economic valuation results, and assay data if anything significant is identified which may lead to development. Community representatives are invited to ICBG annual meetings with assistance for travel and lodging provided.
- A record of the dates and locations of visits, permission granted, people hired, services paid for, gifts and donations made (i.e. all of the above activities) should be kept, dated and signed by the ICBG team leaders. Procedures not achieved should also be noted. These records are reported to the Fogarty International Centre annually.

The detail of this procedure reflects a thorough consideration of local conditions in PNG. According to the project leader, Prof. Barrows, this procedure has been well adhered-to on all aspects except for some of the documentation (the last point).

Consultation with relevant parties

The ICBG researchers have had several years of collaboration with a wide range of relevant stakeholders, going back prior to the 9 years of research activity in the recent ICBG project phases. Initially the ICBG-funded workshops helped draw together various government and non-government actors from DEC, the Attorney General's Office, the Forest Research Institute (FRI), UPNG, NGO partners and others towards the development of what is now the PINBio Act. This demonstrates that from the beginning the ICBG partners have made considerable effort towards ensuring there are transparent ABS procedures in place in PNG and that they comply.

Through their PIC standard operating procedure the researchers engage directly with indigenous and local communities prior to, during, and after collection activities (noted above). Two NGOs, The Nature Conservancy and Conservation Melanesia, have been involved in the collaborations of the ICBG and provide outreach to communities. The ICBG also provides medical outreach as described above.

The ICBG research team have also worked together with the STRI towards the establishment of a 50 ha nature plot in Wanang, PNG in 2008 (involving the New Guinea Binatang Research Centre, the PNG Forest Research Institute, the University of Minnesota, Harvard, the Smithsonian Institution Global Earth Observatories – SIGEO, the Smithsonian Centre for Tropical Forest Science – CTFS and several PNG government agencies). The research team has been working in partnership with the Wanang community landowners towards the long-term protection of the forest and maintenance of subsistence livelihoods. The research represents the first large-scale study of biodiversity and carbon dynamics in PNG, fostering collaborations with local communities and PNG scientists (STRI, 2008).

Benefit-sharing

There are memoranda of understanding dating to 2002, renewed in 2008, between the University of Utah, University of Minnesota and the University of PNG, establishing terms of research collaboration between these parties. The memoranda make commitments towards compliance with the CBD and UN Convention on the Law of the Sea, notably that it is recognised that PNG has jurisdiction over its biological resources. The agreement describes commitments towards improving education and scientific infrastructure in PNG; contributions towards the conservation and sustainable use of biodiversity; transfer of knowledge, expertise and technology related to the collection, storage bioassay-guided isolation and characterisation of natural products and therapeutic agents.

In the event of commercialisation, the agreement specifies joint-promotion of intellectual property and a percentage of total income derived from the commercialisation of transferred biological resources (including income derived from use of resources by third parties under license arrangements).

The terms of collaboration are divided into 2 phases. The first phase deals with initial collection activities and investigations. This phase specifies joint research applications between Utah (and collaborative partners) with UPNG, encouragement of joint fieldwork, analysis by bioassay at both Utah and UPNG and co-authored publications. When conducting taxonomic identification of species, duplicate plant vouchers are deposited with UPNG and the FRI for their own scientific use.

Phase 2 refers to the development of lead compounds and materials identified in phase 1 towards commercial products. The trigger point for phase 2 is the application for patent coverage by Utah and UPNG on compounds or genetic constructs from PNG genetic resources. Separate agreements are made for each product entering phase 2 development, and even for synthetic materials for which the natural product provided a key development lead. The agreement notes that all scientists and individuals who contribute to the identification and discovery of new therapeutics, molecular probes, genetic constructs or pharmaceuticals will be compensated in terms of royalty proceeds arising from patent agreements. Compensation also includes milestone payments at key stages of clinical development. If a natural product is required for sourcing towards commercial agents, then PNG will be the first source of the raw material. Also, if indigenous knowledge is involved in the collection of samples or development of commercial agents, then suitable recognition would be given to their intellectual property, including suitable recompense and patent inventorship where appropriate. The memoranda specify that any licenses granted to agents or companies on any patents arising from this collaboration must abide by the terms of the agreement

(University of Utah-UPNG, 2002; University of Utah, University of Minnesota and UPNG, 2010). Table 9.1 summarises some of the main monetary and non-monetary benefits derived from the ICBG collaboration so far.

Whether or not payments for specific individuals can be considered to be 'benefit-sharing' is debatable. Payment for the contribution of labour can be considered a basic right. However, this project clearly also provides some additional upfront payments that might be considered monetary benefit-sharing, such as the appreciation fee, the donation to a school fund and others.

Impact

Apart from the tangible impacts described above for UPNG student completions and support for UPNG laboratories and herbariums, there are considerable scientific achievements from research conducted at the Wanang nature plot. The pilot Wanang study resulted in 12,136 herbarium specimens deposited in local herbariums representing 5,519 plants and at least 536 species collected and

Table 9.1 Summary of monetary and non-monetary benefits related to the ICBG PNG project

Monetary	Non-monetary
Sustaining contributions to the UPNG herpetarium and National FRI herbarium in Lae.	Education and training: 65 UPNG student projects, 39 of them resulting in completion of degrees or certificates (undergraduate, honours, masters).
Supported collection expeditions, travel costs and field supplies, infrastructure (air conditioning, computers etc).	Workshops and support towards development and implementation of research permits, PIC and benefit-sharing procedures in PNG.
In the past 8 years of activity approximately $350,000 has gone to herbaria activities, students and infrastructure (most to UPNG). Has provided equipment and supplies for the bioassay lab for Honours and Masters students at UPNG pursuing diverse topics.	Expanded opportunities for PNG researchers through access to grants, fieldwork, updated laboratories and equipment.
Considerable milestone payments have been made to UPNG under the MOU (>$100,000) with the most substantial payments made in 2004.	Seventeen co-authored publications (UPNG-Utah/Minnesota), and 3 patents (1 current, 2 pending).
The Wanang nature plot activities provides opportunity for development of PNGs scientific capacity through postgraduate fellowships (funded by an external philanthropic donation of $250,000).	Biodiversity conservation related scientific knowledge as described below.

identified. An online checklist of New Guinea woody plant genera has been produced including 1,073 taxa, their respective families, growth habits, and status as endemic, native, introduced, or cultivated plants (STRI, 2008). Related to these have been a number of ICBG-funded student training/projects on woody plant identification and enumeration, and herbivorous insects, as well as data collection on forest carbon dynamics. These have important implications for applied taxonomic knowledge of various species and ecosystems, thus contributing to the conservation of biodiversity in PNG. Other important medical discoveries of benefit to PNG are described below.

Products of the R&D

There have been a number of products of the many R&D projects undertaken in this ICBG. These include:

1 A predecessor collaboration, between UPNG and a research consortium headed by the University of Utah, funded by the US National Cancer Institute, helped lay the foundation for the ICBG and demonstrated good partner intentions of the University of Utah for the ICBG collaboration (Barrows, pers. comm, 8/6/12). This antecedent work led to the discovery of HTI-286. HTI-286 is a synthetic experimental anticancer drug recently in phase II clinical trials for the treatment of lung cancer (and with potential prostate cancer application). The lead structure for the development of HTI-286 was the sponge tripeptide hemiasterlin, which was collected in PNG under the auspices of the ICBG collaboration (Andersen and Roberge, 2005). Considerable milestone payments have been made to UPNG under the MOU (>$100,000) with the most substantial payments made in 2004. This money was used to build laboratory space on the main Waigani campus of UPNG. HTI-286 has not yet completed clinical trials.

2 A project conducted at the ICBG supported UPNG BioAssay lab and herpetarium has led to the development of new antivenoms for the lethal Papuan taipan snake (*Oxyuranus scutellatus*) (see Figure 9.1). The research was conducted by Mr David Williams from the Australian Venom Research Unit and Mr Owen Paiva, a UPNG student studying his masters. The antivenom has passed pre-clinical toxicity testing, is under GMP production at an institute in Costa Rica and is ready for clinical trials. The new antivenom is reputed to have superior stability, shelf life and is considerably cheaper to produce than existing taipan antivenoms. Following this success, the researchers are working on another antivenom for the small eyed snake (*Micropechis ikaheka*) which will soon be ready for production (University of Utah, 2010).

3 A traditional medicine selected for assessment and development by the Traditional Medicines Taskforce, headed by UPNG researchers, was a topical analgesic and anti-inflammatory preparation manufactured by a PNG woman, Minnie Bate (see Figure 9.2). Ms Bate uses a vegetable oil extract and a species of fresh or dried/stored lichen (subsequently identified as *Parmotrema saccatiboum* and *Pyxine cocoes*) which was pharmacologically examined by UPNG and Utah researchers. The collected data chemically

validated Ms Bates' traditional medicine and has supported her marketing and product distribution to other countries such as Japan. Additionally, the empirical results of the study have led to adjustments in the strength of the preparations to improve efficacy, and reformulation to use coconut oil instead of vegetable oil (University of Utah, 2010).

Discussion

The permitting procedures for access and PIC are extensive and intended to ensure that informed permission is obtained by as many stakeholders as possible. This decreases the likelihood that there will be parties that are excluded or that have objections to the research activities once they have begun. The only evident drawback relating to these access procedures are that they may be quite time consuming.

According to the researchers involved (Prof. Matainaho and Prof. Barrows), and reports in response to queries by the Fogarty International Centre, there have been few issues related to access and PIC. Occasionally documentation of activities in the field has been limited (e.g. missing dates, names of people involved, amounts paid), and on occasions there has been limited information exchange (e.g. explanation of taxonomic findings or similar). These are likely affected by the challenges of working for prolonged periods in remote areas through translators, while trying to meet multiple objectives.

The project takes a comprehensive approach to benefit-sharing at all stages of the R&D process (from access to potential commercialisation). This

Figure 9.1 The UPNG herpetarium which has benefited from the ICBG project

Source: Prof. Louis Barrows, http://www.pharmacy.utah.edu/ICBG/pdf/student_achieve.pdf (Accessed 6/6/12, used with permission).

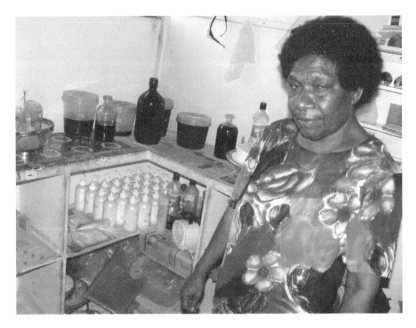

Figure 9.2 Ms Minnie Bate formulating traditional medicines, PNG

Source: Prof. Louis Barrows, http://www.pharmacy.utah.edu/ICBG/pdf/student_achieve.pdf
(Accessed 6/6/12, used with permission).

ensures that there are tangible benefits in terms of training (at UPNG and in communities), small monetary benefits for communities, joint publications and educational opportunities for PNG researchers, among other things. The agreements and procedures ensure benefits for UPNG as the main collaborating institution in PNG, but also local people who are involved in the project. By ensuring benefits are shared at multiple stages of the R&D activities, stakeholder satisfaction is likely to be higher than if the project had adopted solely an 'end product' benefit flow (e.g. a royalty upon commercialisation). Royalties upon commercialisation often do not eventuate unless the bioactive molecules identified as useful are also found to be safe. This can take decades of analysis through clinical trials and considerable investment – often millions of dollars. Minimising community expectations and explaining these costs and timelines also significantly reduces the likelihood that the communities in question will be frustrated by any lack of monetary benefits in the long term.

Note

1 PNG Kina is worth $0.47 (3/9/2012). Weekly minimum wage is 37.4 Kina per week (approximately $17.50). Therefore the rate paid to research assistants by the researchers is

at or above the minimum wage, which is rarely achieved in practice in PNG particularly in rural areas (US State Department, 2009).

References

Andersen, R.J. & Roberge, M. (2005). 'HTI-286, a synthetic analog of the antimitotic natural product hemiasterlin'. In Cragg, G.M, Kingston, D.G.I. and Newman, D.J. *Anticancer Agents from Natural Products*. CRC Press, Boca Raton FL, pp. 267–280.

Cragg, G.M., Katz, F., Newman, D.J. & Rosenthal, J. (2012) 'The impact of the United Nations Convention on Biological Diversity on natural products research'. *Natural product reports,* 29(12), 1407–1423.

NIH, (2012) '*Our Mission and Vision*'. Available at: http://www.fic.nih.gov/About/Pages/mission-vision.aspx, accessed 6/6/12.

STRI. (2008) '*Establishment of the Wanang CTFS/SIGEO Plot. CTFS/SIGEO to study forests of Papua New Guinea*' October 20. Available at: http://www.stri.org/english/about_stri/headline_news/news/article.php?id=884, accessed 6/6/12.

University of Utah. (2010) '*ICBG PNG: Conservation and sustainable use of biodiversity in PNG*'. Available at: http://www.pharmacy.utah.edu/ICBG/, accessed 9/5/12.

University of Utah and UPNG. (2002) Memorandum of Understanding. Focus: A Scientific Research Collaboration to Investigate the Biological, Chemical, and Medicinal Properties of the Biodiversity of Papua, New Guinea and to Establish Economic Value Thereof.

University of Utah, University of Minnesota and UPNG. (2010) Memorandum of Understanding. Focus: A Scientific Research Collaboration to Investigate the Biological, Chemical, and Medicinal Properties of the Biodiversity of Papua, New Guinea and to Establish Economic Value Thereof.

US State Department. (2009) '*2008 Human Rights Report: Papua New Guinea*'. Available at: http://www.state.gov/j/drl/rls/hrrpt/2008/eap/119053.htm, accessed 3/9/2012.

10 The ICBG Panama project

Background

There have been several papers written describing the approach of the research team working on the Panama ICBG, as a bioprospecting project framed quite broadly so as to contribute to conservation and development as per the intent of the CBD ABS articles. Most of these have been written by the team involved or by others at the Fogarty International Centre with a role in the ICBG grant program, as well as a handful of other papers reviewing the project. Therefore, it is worth examining the case relatively briefly via literature review, in lieu of actual fieldwork in Panama (some comments from the project team and FIC staff were obtained for this ICBG project case study).

As described by several members of the project team, the Panama ICBG is a drug discovery effort that emphasises local benefit. Their team acknowledges that it is not realistic for host countries or local communities to expect royalties, given the incredibly small chance of discovering a marketable drug. Rather, the ICBG was established so as to concentrate as much of the drug discovery process in Panamanian laboratories, providing training and skills development, local employment, enhanced facilities and other benefits (Kursar et al., 2007). The project has been funded as part of the ICBG since 1998 and consists of collections, assaying of samples for activity against specific disease targets and the purification of active compounds. The project just completed its third phase of ICBG funding in 2013 and has applied for a 4th phase in 2014. The collections program is based at the Smithsonian Tropical Research Institute (STRI), the University of Panamá and the Panama's Institute of Advanced Scientific Research and High Technology Services (INDICASAT), all in Panama City, and has focused on plants, endophytic fungi, marine bacteria and macroalgae, corals and sponges, most of which are collected from Panama's protected areas (Kursar et al., 2007). Bioassays are primarily carried out at Panama's national laboratories at INDICASAT, with some anti-bacterial bioassays conducted at STRI and at UC Santa Cruz. Active components are purified in laboratories in both Panama and the United States, including UC San Diego, Oregon State University, University of Connecticut and Duquesne University (Kursar et al., 2007). The team utilises the ecological knowledge of STRI to understand chemical defences and to utilise ecological criteria for collection strategies for targeted drug discovery (Coley et al., 2003; Kursar et al., 1999; ICBG, 2012).

At an earlier funding stage, the project was partnered with Novartis Institutes for Biomedical Research, who provided $50,000 for laboratory equipment for use in Panama (Dalton, 2004). Recently, the project has also attracted industry funding from Eisai Inc. and Dow Agro-sciences who contribute annually to support research in Panama. Additionally, both have been active partners in research, contributing expertise and sample analyses and hosting Panamanian students and scientists at their institutions (Coley and Kursar, pers. comm. 6/4/2014).

Access and R&D

Relatively recently, the Executive Decree 25 of Panama, dated 29 April 2009 has updated its ABS regulations (under the General Law of Environment of Panama). For this purpose a special unit has been created – the 'Unidad Nacional de Recursos Genéticos' (UNARGEN) under the supervision of the National Environmental Authority. According to these regulations, collection and permits issued by UNARGEN are required for research within the country and export permits are required for sending samples out of Panama (Gupta, c2011).

The ICBG requires memoranda of understanding between the partners of all its projects, and thus the Panama project has a similar agreement to that already discussed regarding the Madagascar and PNG ICBG projects. These 2-stage agreements emphasise some upfront benefits (monetary and/or non-monetary) for academic or pre-commercial research, with a requirement that additional negotiation of benefit-sharing is required, typically milestone payments and royalties, at the stages of commercial development and marketing of new drugs or products (see Kursar et al., 2006). This contract regulates the ABS aspects of the partnership, which also has to comply with relevant laws of Panama.

The project has focused on building research capacity in Panama, through training of young scientists, infrastructure enhancement and support for senior scientists to conduct research. Plants and now primarily micro-organisms are collected in marine and terrestrial protected areas. Extracts are screened in the therapeutic areas of cancer, neuroscience, antibiotics and neglected tropical diseases (e.g., malaria, leishmania and Chagas' disease), as well as agriculture (ICBG, accessed 25/3/14). The ICBG has supported the enhancement and in many cases the establishment of new research laboratories in Panama, including microbiology, natural products and disease bioassay laboratories (Coley and Kursar, pers. comm. 6/4/2014).

The research team highlight a number of important outcomes from the R&D:

- The identification of anti-leishmanial alkaloid compounds from *Guatteria* species which have a 65-fold higher toxicity towards cells of disease organisms than towards human macrophages;

- The discovery of a novel cysteine protease inhibitor, gallinamide A, from a marine cyanobacterium with very promising activity in malaria and cancer;
- Discovery of a novel cyclic peptide, coibamide A, from a marine cyanobacterium collected from the National Park and UNESCO World Heritage Site of Coiba Island, which has a novel mechanism of killing cancer cells;
- Discovery of a novel histone deacetylase inhibitor, santacruzamate, from a marine cyanobacterium with very promising activity in cancer;
- The first cytotoxicity (or whole cell) anticancer assay in Panama, including the isolation of 40 anticancer compounds, 13 of which are new to science, from the purification of active molecules from plants;
- The development of a series of assays to test for activity against leishmaniasis, malaria and Chagas' disease – neglected diseases especially relevant in the 'developing world';
- The development of an anti-malarial assay using a fluorescence probe by INDICASAT, which is of great practical relevance for laboratories in the 'developing world', given the many issues of using traditional anti-malarial assays that use radioactive materials. This benefit was also noted by interviewees from Madagascar, who had received training on the use of the assay in Panama (Kursar et al., 2007, pp. 2794–2795; Coley and Kursar, pers. comm. 6/4/2014).

Further to this, the following bullet points provide a few highlights that showcase results of this past 5.5 year period of ICBG support (phase 3, 2008–2013 inclusive):

- A large number of habitats have been accessed for their unique microorganisms, including terrestrial and marine cyanobacteria, lowland jungle and cloud forest endophytic fungi, aquatic fungi, heterotrophic bacteria from sediments and marine and terrestrial symbiosis involving invertebrates, vertebrates and insects, resulting in 5-year total of thousands of strains;
- The isolation and creation of a microbial culture bank at the INDICASAT laboratories, currently holding more than 5,000 isolates;
- Development of innovative biological assays, including BioMap and cytological profiling;
- A similarly large number of extracts have been screened for antiparasitic (malaria, leishmania and Chagas' disease), cancer cell cytotoxicity (H-460, MCF-7), antibiotic (BioMap), neurochemical, anti-inflammatory and agrochemical screens, and many screening hits were obtained;
- Discovery that the extracts of endophytic fungi from cloud forest plants have a higher incidence of biological activity than do those isolated from lowland rain forest plants;
- More than 15 novel natural products were isolated and characterised from marine bacteria and corals;
- More than 40 natural products were isolated and characterised from endophytic fungi. Eighteen of these are novel, including the very potent

coibanoles and mycoleptodiscins. These are from two fungi, both isolated from the plant *Desmotes incomparibilis*, a genus endemic to Coiba;

- Approximately 50 compounds were isolated and characterised from marine cyanobacteria, and these include the anti-inflammatory coibacins, HDAC inhibitor santacruzamate, antimalarial bastimolides A-C and antiproliferative metabolites veraguamides;
- More than 34 almiramide analogues were synthesised and screened against *L. donovani* and *Trypanosoma brucei*, and the target in the latter parasite identified as the glycosome, an organelle specific to kinetoplastid parasites;
- Extensive synthetic efforts on the anticancer lead coibamide A were conducted, and have revealed insightful SAR features in this molecular class, as well as mechanism of action and in vivo studies;
- Application of molecular networking based on MS2 data to profile the metabolomes of diverse marine and terrestrial bacteria. This technology has been transferred to Panama through one of our most recent US postdoctoral scientists that has been stationed in Panama, Dr Amanda Fenner;
- Synthesis and semi-synthesis of diverse natural products (NPs), analogs and fragments in several compound series, including a chiral synthesis of DHOYA (a NP fragment), palmarumycin-CP18, almiramide, and santacruzamate;
- Meta-genome sequencing of several marine cyanobacteria consortia from Panama with resulting deep insights into the genes and mechanisms of natural product biosynthesis;
- The widespread application of molecular-based phylogenetic analyses of endophytic fungi, bacteria and cyanobacteria, and as a result, many new lineages, species and genera were identified, such as the new genus of cyanobacterium, *Okeania*;
- Thousands of plant samples were collected from various locations around Panama, including the Chucanti Reserve in the Darien Province, the Santa Fe National Park, Escudo de Veraguas Island, and the Coiba Park Archipelago. Many of these represent new distribution records, and several are new to science;
- In 2011, Alicia Ibanez published the authoritative and definitive guide to the plants of the Coiba National Park (Guia Botanica del Parque National Coiba);
- Continued advisory roles to the Coiba Scientific Committee on underpinning scientific issues of importance to conservation activities in the National Park;
- Training of hundreds of Panamanian students at the high school, undergraduate, graduate and postdoctoral levels (Coley and Kursar, pers. comm. 6/4/2014).

Notably, many of these developments have important implications for the conservation of biodiversity. The identification of new microbes, plants, publication of field guides and training of local scientists and students all make some contribution.

Benefit-sharing

The split of revenue from potential commercial benefit has been praised by Scholz (2004, p. 230), who noted that:

> ... the [principal investigators] PIs in the Panama ICBG project designed a "club model" that is notable for its unusually favorable treatment of the host country, local knowledge and the conservation agenda.

All incoming monies are split such that 50 per cent goes to the institution in Panama which makes the discovery and 50 per cent to the club of collaborators, who get equal shares. This club includes STRI, the University of Panama, INDICASAT and Panama's interior department (ANAM) (Coley and Kursar, pers. comm., 6/4/2014).

This is a very high split for the host country compared with some other examples discussed so far in this book. Scholz (2004; and Kursar et al., 1999) also noted that in the future, if indigenous groups who have worked with the ICBG choose to participate (to date they have opted not to be part of drug discovery activities, but rather have participated only on an inventory of medicinal plants), they can be added to this 'club' and the shares are again apportioned equally. Papers by the project investigators do not emphasise this aspect of benefit-sharing, due to its low likelihood (Kursar et al., 2006, 2007). This is an important consideration, because past bioprospecting activities have sometimes failed to temper local expectations regarding the likelihood of long-term payments (e.g. the Falealupo case in Samoa, and the ICBG Peru – see Greene, 2004).

The foremost benefit highlighted by members of the project has been the training provided to scientists in Panama. In the first 8 years of the program, there have been 70 scientists involved, 22 of whom have gone on to complete higher degrees or postdoctoral training overseas (Kursar et al., 2007, p. 2795). Notably, Panama had, until recently, no domestic PhD programs at its universities (Burtis, 2007). In addition, they noted the purchase of laboratory equipment and instrumentation for laboratories in-country, so that much of the research conducted by the ICBG project is conducted in Panama.

Kursar et al. (2007) noted that the training, infrastructure improvements and research of the project are produced with modest funding of about $500,000 per year (now closer to $800,000 per year). This limits the amount of benefit-sharing possible or likely from the project, particularly in terms of upfront payments (as in the ICBG Madagascar project). This 'lack of scale' is criticised by Burtis (2007) as insufficient when compared with the research budgets of commercial biotechnology and pharmaceutical companies. However, partnerships with Eisai, Inc., and Dow Agrosciences contribute annually to support research in Panama. Additionally, both have been active partners in research, contributing expertise and sample analyses and hosting Panamanian students and scientists at their institutions. Should potentially commercial discoveries be made, there is the possibility of milestone and royalty payment benefits (as has occurred in the ICBG Madagascar) (Coley and Kursar, pers. comm. 6/4/2014).

Impacts

The major positive impact from the project is towards building a drug discovery research program in Panama. A second goal has been to integrate scientific research with conservation. The many Panamanian scientists involved in the ICBG recognise the value of biodiversity, and also have access to the government and policy makers. This strong urban voice of Panamanian scientists has been instrumental in a number of policy decisions. One has been helping to shape the laws regarding ABS such that research is facilitated and benefits are transparent and fair (Coley and Kursar, pers. comm. 6/4/2014). As Joshua Rosenthal of the FIC notes:

> It is probably worth emphasizing that the training is not just lab techs, but they have done a lot of long term training of students, through post docs and professors and established several laboratories. A parallel pipeline of young U.S. Scientists has grown up together with Panamanian investigators in at least three institutions, and this bodes well for long term collaborations.
>
> (Rosenthal, pers. comm. 27/3/2014)

Adding to this conversation, Phyllis Coley from the project team (University of Utah) explains:

> . . . we bring a US postdoc to Panama for at least a year-long internship. This helps with transfer of technology from north to South, is a life changing experience for the postdocs, helps them understand international collaborations, makes them more competitive in getting a University position, and establishes a lasting connection between Panamanian and US labs.
>
> (pers. comm. 6/4/2014)

Alongside training, it is also worth noting that this group has produced over 90 scientific publications over the 15 years of support, which Flora Katz of the FIC indicates makes 'a significant contribution to the scientific literature on natural products, biodiversity and conservation in Panama' (Katz, pers. comm. 1/4/2014).

Another significant and relevant impact in the CBD ABS context that has occurred from the Panama ICBG is the establishment of a National Park on Coiba Island. Based in part on the research conducted on the Island, the Panamanian Government have the area National Park status in 2004, and subsequently it also received UNESCO World Heritage status in 2005 (including the Park and surrounding marine area) (Cragg et al., 2012). Staff and affiliates of the ICBG indicate that this is likely to benefit Panama in terms of increased tourism and the resultant economic benefits that arise from tourism, as well as for conservation and further research (Cragg et al., 2012).

Another useful side-benefit of the project has been the creation of a detailed natural products discovery database by a Panamanian company that is accessible to research institutions in Panama. As Cragg et al. (2012) noted all ICBGs are required to track collections, bioassays and chemistry and so there is a complete record of all materials and research activities. These authors also noted that this database will help Panama comply with the Nagoya Protocol requirements for monitoring the utilisation of genetic resources and information sharing with the CBD Clearing House mechanism (Cragg et al., 2012).

Discussion

The positive aspects of this ICBG project have been described in detail in several articles, including those written by project staff and Fogarty International Centre authors. This and the fact that it has received grant funding in 3 phases so far (15 years) suggest that this is regarded as an exemplar case. The R&D outcomes, tangible benefits in terms of training for Panamanian researchers, as well as the assistance in establishing a National Park which has now also attracted UNESCO World Heritage status, are all important achievements. These outcomes also imply that the project has been designed in a broad enough way so as to encompass conservation and development benefits for the host country. Further evidence of the 'robustness' of the model they have built over the years of the project is the award of the first ever Global Environment Facility Nagoya Protocol Implementation Fund to the Panama ICBG Consortium and Panama Government (Cragg et al., 2012). This grant is intended to support projects which implement 'ABS' through biodiscovery projects, and to assist with ratification and implementation of the Nagoya Protocol in the host country. The split of revenue from potential commercialisation is generous, and the ability to include interested indigenous groups at a later date is also a novel and encouraging inclusion. Despite some earlier speculation that the Panama project may collapse due to a lack of involvement of indigenous communities (see Scholz, 2004, p. 232), the project investigators seem to have achieved a balance – it is difficult to suggest otherwise without conducting interviews on the ground. Many of these elements suggest that the project investigators have learnt from some of the more controversial experiences surrounding ICBG projects, including the ICBG Maya (see Cragg et al., 2012) and the ICBG Peru (see Greene, 2004). Indeed, in both of those ICBGs many of the issues stemmed from failures to establish agreements with indigenous populations, in circumstances where 'appropriate indigenous authority' is not vested with any one specific representative body or group, but rather there are many indigenous voices. The lack of 'utilisation of traditional knowledge associated with genetic resources' in both the Panama and Madagascar ICBGs is therefore not particularly surprising given the concerns raised by indigenous groups and NGOs in these earlier ICBG projects.

However, the Panama project has not been without its limitations. As Dalton (2004; and Kursar et al., 2007) noted, many new compounds have been identified, some have been evaluated as pre-clinical leads, however, none have so far reached clinical trials, despite several million dollars of investment and years of research to date. The project team indicate that retrospective studies of the drug discovery process suggests that it takes an average of 15 years from discovery of a new drug lead to its introduction into the market place, and that someplace around 15,000 compounds on average must be assayed for activity to find one which will go all the way to drug status. Thus, it is premature to use the metric of number of compounds reaching clinical trial as a measure of the success of this ICBG program, or really any of the ICBG programs to date (Coley and Kursar, pers. comm. 6/4/2014). As noted by Burtis (2007), there has been a lack of industry investment in the project (until recently). Dalton (2004) argued that the lack of interest from industry is at least in part due to a negative image of bioprospecting, and due to concerns about benefit-sharing and legal certainty. His suggestion of certificates of origin has been discussed for several years and has now essentially been incorporated into the operation of the Nagoya Protocol, including through the operation of the ABS Clearing House mechanism.

While the Clearing House is not yet established – it has been discussed at the Open-ended Ad Hoc Intergovernmental Committee for the Nagoya Protocol on Access to Genetic Resources and the Fair and Equitable Sharing of Benefits Arising from their Utilization (ICNP), where negotiators are narrowing down the bracketed text on its operation. In essence, it is expected that national ABS systems will require researchers to track their extraction and use of genetic resources and establish national checkpoints to monitor transfers of the resources at 'any stage of research, development, innovation, pre-commercialisation or commercialisation' (Article 17.1(a)(iv) of the Nagoya Protocol). The national system will also require the access permit, or an equivalent to be sent to the ABS Clearing House which will be internationally recognised as a 'certificate of compliance'. This process would then be enforced through compliance measures for users conducting R&D within the country's jurisdiction (Articles 15 and 16). What is not so clear, is regarding what legal recourse would be available for users who do not comply with the provisions of the access permits and MAT of other countries. For example, non-Parties to the Nagoya Protocol would not necessarily have such measures in place for compliance (e.g. the United States is expected by many to remain only a signatory to the CBD, and is therefore unable to ratify the Nagoya Protocol). This then leaves avenues for legal recourse to contract law (if there is a MAT contract), meaning that, at a minimum these contracts need to clearly state the dispute settlement process as well as the jurisdiction through which disputes would be resolved.

The ICBG program is a unique case, given that the grant funding comes from the US Government. As a signatory to the CBD, the US Government is not bound by the obligations of the CBD until it has ratified (accepted or approved) it. However, the ICBG operates such that it follows CBD-compliant principles (as described on the ICBG website). In the near future, it will be interesting to see how the ICBG program would comply with the national ABS requirements of countries as they implement the Nagoya Protocol, and if any changes in their procedures will be required.

References

Burtis, P. (2007) 'Can bioprospecting save itself?' *Journal of Sustainable Forestry,* 25(3–4), 218–245.

Coley, P.D., Heller, M.V., Aizprua, R., Araúz, B., Flores, N., Correa, M., Gupta, M.P., Solis, P.N., Ortega-Barria, E., Romero, L.I., Gomez, B., Ramos, M., Cubilla-Rios, L., Capson, T.L. and Kursar, T.A. (2003) 'Using ecological criteria to design plant collection strategies for drug discovery'. *Frontiers in Ecology and the Environment,* 1(8), 421–428.

Cragg, G.M., Katz, F., Newman, D.J. and Rosenthal, J. (2012) 'The impact of the United Nations Convention on Biological Diversity on natural products research'. *Natural Product Reports,* 29(12), 1407–1423.

Dalton, R. (2004) 'Natural resources: Bioprospects less than golden'. *Nature,* 429(6992), 598–600.

Greene, S. (2004) 'Indigenous PEOPLE Incorporated? Culture as politics, culture as property in pharmaceutical bioprospecting'. *Current Anthropology,* 45(2), 211–237.

Gupta, M.P. (2011) 'Access and benefit-sharing: Viewpoints from provider countries'. Available at: International Council for Science website: http://www.icsu.org/freedom-responsibility/advisory-documentation/pdf-images/ABS_WS_Presentation_Gupta.pdf, accessed 7/4/2014.

ICBG. (2012) Panama ICBG. webpage: http://www.icbg.org/groups/panama.php, accessed 25/3/14.

Kursar, T.A., Caballero-George, C.C., Capson, T.L., Cubilla-Rios, L., Gerwick, W.H., Heller, M.V., Ibañez, A., Linington, R.G., McPhail, K.L., Ortega-Barria, E., Romero, L.I. and Coley, P.D. (2007) 'Linking bioprospecting with sustainable development and conservation: The Panama case'. *Biodiversity and Conservation,* 16(10), 2789–2800.

Kursar, T.A., Caballero-George, C.C., Capson, T.L., Cubilla-Rios, L., Gerwick, W.H., Gupta, M.P., Ibañez, A., Linington, R.G., McPhail, K.L. Ortega-Barria, E., Romero, L.I., Coley, P.D. and Solis, P.N. (2006) 'Securing economic benefits and promoting conservation through bioprospecting'. *Bio Science,* 56(12), 1005–1012.

Kursar, T.A., Capson, T.L., Coley, P.D., Corley, D.G., Gupta, M.P., Harrison, L.A., Ortega-Barría, E. and Windsor, D.M. (1999) 'Ecologically guided bioprospecting in Panama'. *Pharmaceutical Biology,* 37(supplement), 114–126.

Scholz, A. (2004) 'Merchants of diversity: Scientists as traffickers of plants and institutions'. In Jasanoff, S. and Martello, M.L. (eds) *Earthly Politics: Local and Global in Environmental Governance,* MIT Press, Cambridge, MA, pp. 217–238.

Part III

Overview, discussions and themes

11 Discussions and themes

This chapter provides some overview and comparison of the cases. The first section draws us back to the objectives of the CBD and the Nagoya Protocol and considers the role played by the case studies that have been examined, and by other potential ABS agreements in the future. Then, rather than re-examining in too much detail the discussion already made in each of the case studies, this chapter seeks to further explore some of the most important themes or issues that arise in biodiscovery activities and from ABS agreements. This will be done with reference to recent developments in the literature on ABS and related areas, national systems and specific ABS agreements.

Overview of the case studies

The analysis of the case studies in this book has been qualitative by necessity. There are no fixed indicators for evaluating ABS projects, but rather there are principles and objectives under the CBD. For the sake of analysis and overview of all of the cases, it is worth returning to the core objectives of the CBD and Nagoya Protocol and the principles underlying ABS. The objectives of the Nagoya Protocol are essentially a re-statement of the CBD objectives:

> The objective of this Protocol is the fair and equitable sharing of the benefits arising from the utilization of genetic resources, including by appropriate access to genetic resources and by appropriate transfer of relevant technologies, taking into account all rights over those resources and to technologies, and by appropriate funding, thereby contributing to the conservation of biological diversity and the sustainable use of its components.
>
> (Article 1, Nagoya Protocol)

Arguably, the original intention of ABS is for the conservation of biological diversity and the sustainable use of its components (the final part of the objective above). However, it has been suggested by others that it has become more about questions of fairness and equity, avoidance of 'biopiracy' and clarification of ownership of genetic resources and associated traditional knowledge (TK) (see Chiarolla et al., 2013; Tobin, 2013). While the latter is something rather immeasurable (we return to that question further on), we might be able to gain some insight from specific cases of the direct contribution of bioprospecting to conservation. Table 11.1 comments on some of the specific gains for biodiversity conservation, obtained through each of the biodiscovery ABS cases we have discussed in this book.

Table 11.1 highlights that bioprospecting can indeed provide direct and tangible benefits towards conservation, even with quite limited budgets (as has been noted in the ICBG cases). The ICBG projects, because of the way the

Table 11.1 The ABS cases – what direct contributions to biodiversity conservation

Biodiscovery ABS case	Main direct contributions to biodiversity conservation
ICBG Madagascar	- Identification of rare and endangered species and detailed taxonomic inventories of several sites across Madagascar. - The above inventories and fieldwork have assisted in the establishment of several protected areas, including Ibity Massif protected area, Montagne des Français temporary protected area, Oronjia forest community protected area, the Ambodivahibe Bay marine reserve, Ankafobe community protected forest. - Continued monitoring of the above protected areas. - Training of local scientists in conservation-related fields (e.g. botany). - Training of local people and conservation projects in local communities (many related to the above protected areas).
Samoan mamala	- Establishment of the Falealupo forest reserve and protection for 50 years (if not longer). - Establishment of ecotourism activities and educational benefits related to the forest and its conservation.
Moroccan argan	- Basic training about forest protection. - Direct incentive to conserve the forest (as source of income and ecosystem services).
ICBG PNG	- Establishment of Wanang nature reserve and a variety of conservation research activities at the plot. - Training of local people and scientists in fields such as botany, contributing to local conservation. - Taxonomic inventory and identification of new woody plant species.
Expedition of Santo	- Extensive inventory of species on the Island of Espiritu Santo. - Educational benefits for local communities related to conservation.
ICBG Panama	- Contributions to the establishment of Coiba Island National Park and UNESCO site. - Training of local people and scientists in fields such as ecology, contributing to local conservation. - Taxonomic inventory and identification of plants new to science.

grant scheme is set up with a secondary aim of biodiversity conservation, have probably been the most successful contributors to this. Their contributions are notable particularly because of the establishment of nature reserves and protected areas, training in fields directly related to biodiversity conservation, and through detailed inventory and identification of species. Some of the cases in the book are probably better described as contributing to 'sustainable use' of biodiversity. For example, the Moroccan argan case in particular sees benefits go towards cooperatives who have existing knowledge of the need to conserve the forest, and their activity towards the production of argan oil and argan-based cosmetic products arguably encourages them to sustainably use biodiversity. Indirectly there has been much activity towards the conservation and restoration of the argan forest by other parties (e.g. the government, NGOs and academic research) because of the need to sustain the forest for the sake of the local communities as well as for its ecological benefits and cultural significance.

What is more common in all of these cases is the contribution towards local community development. If this were tabulated in the same way as the above table, it would go for several pages. This is likely to have occurred because it is often simpler, more cost-effective and the impact is more directly measurable to provide say a school to a community, than it is to establish and fund conservation measures. For example, the ICBG Madagascar case has projects that contribute to both conservation and local community development. It is quite easy to see the benefits for a discrete number of students in a newly built school, or for residents who have received a well. Indeed, because of immediate community needs and priorities, these were chosen. Preventing fire which is threatening ecosystems in Ibity and Firaranzana in central Madagascar is arguably a much more challenging task requiring long-term commitments and monitoring. Similar things can be said about the Moroccan argan case – the cooperatives have voted on immediate benefits for their families and communities. Through their avoidance of harm to the argan forest, which has been part of customary law for centuries, they contribute to its sustainable use and conservation. However, there are other external impacts on the forest that is degrading it, for which other actors have taken up active re-forestation activities. This re-forestation has been a slow and technical process that is probably beyond the capacities of the argan cooperatives, but which has been taken up by NGOs and the government in Morocco.

Although local community development is not explicitly a goal of the Nagoya Protocol or CBD, Article 8(j) of the CBD refers to support for indigenous and local communities which retain knowledge, innovations and practices relevant for conservation and sustainable use of biodiversity. In addition, the Nagoya Protocol now encourages benefit-sharing with traditional-knowledge holders where that knowledge is 'utilised' for R&D towards biodiscovery (Article 5.5). In this regard, several of the above cases are arguably 'ahead of the pack' in implementing benefit-sharing with local stewards of ecosystems that were the source of genetic resources and/or local knowledge providers (e.g. ICBG Madagascar, the Moroccan Argan case, ICBG PNG). The

Samoan mamala case even occurred before the CBD; however, many of the philanthropic benefits received were after the CBD, and thus Paul Cox notes that the original agreement was not explicitly an ABS agreement in the current sense.

Most of the cases in this book are quite exemplary because there are relatively few agreements that share benefits towards conservation or local communities in recent decades since the concept came to prominence. The 2 cases that arise *de rigueur* are the Costa Rica INBio Agreement with Merck & Co and the Hoodia case in southern Africa involving the San people. While these 2 cases have their merits, they have also been criticised on a number of points. The Hoodia case was a retrospective benefit-sharing agreement, for example, and has yielded very limited monetary benefits for the San people in terms of small milestone payments to date, while the plant itself has subsequently been degraded in its arid natural ecosystems by illegal poachers capitalising on the hype surrounding the plant (see Robinson, 2010). The Cost Rica INBio case is often criticised for not resulting in commercially viable drugs, and there have been questions raised about the lack of involvement of local and indigenous people. On the other hand, it has been pointed out that INBio's agreements with Merck and several other partners since have only occurred in National Parks and that Costa Rica is a unique case because 'protected wild lands in Costa Rica have no inhabitants, local farmers or indigenous people' (Gamez, 2007). We can then ask if INBio is guilty of anything other than avoidance of the potential complexities associated with negotiating access and MAT with indigenous and local communities? Indeed this has probably already become a common trend in bioprospecting activities where researchers and institutions prefer the legal certainty of acquiring genetic resources from genebanks, repositories or other institutions (I return to this discussion later on in the Conclusions). Indeed some of the recent ICBG projects appear to be focusing away from utilisation of ethnobotanical knowledge (and much natural products discovery R&D activity is now focused on microbes for which there is often limited TK). On the other hand, the ICBG Panama allows for an indigenous group – the Naso – to 'opt in' and become involved in biodiscovery activities at a later stage, and the ICBG PNG actively engages with local communities as the providers of genetic resources for which the community receives up-front benefits and potential future benefits.

Despite these concerns surrounding INBio, there are still arguments regarding the positive benefits of the INBio model of ABS since it was established and utilised for dozens of agreements:

> Over US$600,000 corresponding to 10 per cent of the research budgets went directly to conservation activities carried out by MINAE [Ministry of the Environment and Energy, Costa Rica]. A significant contribution of more than US$2 million in total, corresponding to research expenditures (salaries, equipment, infrastructures, laboratory supplies, etc.) was transferred to MINAE's Guanacaste Conservation Area, to the University of Costa Rica and to the National University of Costa Rica . . . [and]

approximately US$0.6 million in milestone payments have been shared 50:50 with MINAE [and INBio].

(Gamez, 2007, p. 86)

This makes the INBio model comparable to the ICBG projects in basic quantitative terms (money committed towards conservation activities), when viewed over a series of actual partnerships. However, more detailed analysis of the conservation impacts on the ground would have to be made to determine the extent to which INBio has been able to contribute to conservation and how. What can be said from analysing the cases in this book is that the proportion of money spent on conservation as a component of the overall project in most of these cases, and as a proportion of the research budgets of the companies involved in the projects, is relatively small. For example, in the margins of a conference where I presented on the Moroccan argan case, I was asked if L'Oreal has similar ABS-type arrangements for supply of biological resource derivatives (where R&D was conducted), or if this was just one out of the many likely biological resource supply chains that contribute to the many products sold by the company. A cynical interpretation of this question is that the case was being perceived as tokenism and their marketing was capitalising on that specific case. However, as discussed in the chapter on argan, there is a sense that the Nagoya Protocol has changed the expectations for the cosmetics industry because of its inclusion of 'biochemical compound' derivatives within the definition of 'utilisation of genetic resources'. Therefore this case might be seen as a test case for the industry that will have to navigate the difference between 'biotrade' supply of biological resources and 'utilisation', and which will no doubt be monitored by the NGOs and ABS observers interested in this issue (discussed further below).

These reflections also raise the question of whether it is worth having such an elaborate system just to contribute in such a minor way to conservation? I have heard several commentators in side-events of the CBD COP suggest that the money spent on ABS systems, negotiations and policy-making could be better spent directly on practical conservation activities – enforcing protected areas, eradicating exotic species, monitoring vulnerable and threatened ecological communities and understanding key pressures causing habitat degradation. However, for the proponents of ABS the presumption is that this money would not make its way to conservation without the concept of ABS – it would be spent on R&D for drugs via HTS or for some other purpose. Indeed it can be seen in the ICBG projects that synergies are able to be found between species inventory of an area with conservation objectives and utilisation of that inventory to determine potential novel drug leads. Further, it might also be emphasised that drug discovery itself is a socially beneficial activity (which often seems forgotten in ABS policy-making and in academic articles on ABS). What is important to consider in the drug discovery process is whether the investigations are targeting cures for tropical and neglected diseases affecting the developing world, or whether they focus heavily on lifestyle diseases such as cancer and heart disease that are of greater concern in the developed world (thus perpetuating colonial disparities). In this book, we have seen a mixture, with some promising developments from R&D in the ICBG projects particularly relevant for developing countries.

The other glaring factor to be considered when thinking '*is an ABS system worth it?*' is regarding ideas of 'fairness and equity' and 'rights over those resources' which we see in the aforementioned objective of the Nagoya Protocol. In fact, it would be very naïve to think that the Nagoya Protocol has emerged because of a singular belief that bioprospecting will make great contributions to conservation of biodiversity. Rather, it is the result of the pursuit of a more fair and equitable process which respects countries sovereign rights over biodiversity (particularly biodiverse developing and middle income countries) and which respects the rights of indigenous and local communities. Certain actors and negotiators to the Nagoya Protocol were able to capitalise on the advancements made with the development of the UNDRIP, seeking to advance the rights of indigenous and local communities with regards to their stewardship, conservation and control over of genetic resources, and also relating to the protection of TK that might be sought and utilised by others. While there are administrative costs and burdens for governments in developing and implementing an ABS system, there are also intangible and unquantifiable benefits relating to the assertion of these rights. As many countries now consider the ratification of the Nagoya Protocol, they will often be making these evaluations – often based on past experience with research projects and bioprospecting activities. The negative experiences, discussed in Robinson (2010), will have to be weighed against the perceived and potential benefits. The likelihood of these benefits – often small up-front but with a very small chance of more sizable benefits later – must also be considered.

To explore these potentials in more detail it is worth examining a number of relevant themes and ABS principles. For example, the concept of prior informed consent (PIC) is central to the ABS process and arguably provides indirect benefits regarding transparency of entry to and from countries, sites and for the transfer of genetic resources. Other key principles such as mutually agreed terms (MAT), ideas such as 'fairness and equity', and also factors such as widely held versus locally held TK are also worth examining.

Permits and PIC

PIC for access to 'sovereign' genetic resources of a country is typically achieved through a permit process. There may also be a requirement to obtain the PIC of local access providers, whether we are talking about genetic resources, associated TK or both.

One of the most likely ways that Parties to the CBD can control research access is through a permit system. Permits can be required as part of the visa process for entry into a country or they can be applied separately (e.g. Australian residents must also apply for research permits for access to Australian genetic resources in certain jurisdictions). For the implementation of the Nagoya Protocol, it makes sense for countries to at least link permits with visa control, so that customs and border authorities can check samples that might be exported. While this is often required for phytosanitary or quarantine reasons, these authorities may not be aware of research access conditions unless there is a system put in place and/or lines of direct communication between the access permit granting body (the National Competent Authority for ABS or a delegated authority) and the customs authority. Indeed this might be a compliance checkpoint for countries to check the outgoing genetic resources of users and incoming genetic resources of users (whether foreign or domestic citizens). However, it is worth noting that customs control is likely to be an incredibly difficult way to enforce genetic resource transfers, given low capacity in many countries, porous borders, marine borders, transferability of genetic resources and often low likelihood of endemic species in any case (consider a country like Indonesia, for example, which has 17–18,000 islands!).

Some countries may require more than one permit, in addition to an entry visa. For example, Vanuatu requires a research permit for research on biodiversity from the Department of Environment and Conservation. It also requires a separate permit for research on Vanuatu's culture, TK, folklore, and practices from the Vanuatu Cultural Centre. While coupling an entry visa with requirement for a permit may seem an effective way to discourage misappropriations of genetic resources – 'no permit no entry' – it is probably easily circumvented by entering as a tourist. Reliance on this might also fail to account for that fact that citizens of that country could also seek to conduct R&D on biological resources on genetic resources. A good example of this is discussed in Robinson (2012) regarding the Cook Islands whereby a citizen (who was not required to obtain a research permit) established an ABS agreement with the Koutu Nui – an indigenous representative body. Because of the potential to circumvent permits by say a researcher entering and posing as a tourist, the CBD has opted for a Clearing House which would display legally obtained permits, whereupon potential infringements of provider country permits might be checked at a later date and punished under the laws of the user country (discussed further in monitoring and compliance).

In several of the cases discussed in this book, there were explicit permissions obtained from the provider governments in question. In others, PIC agreements were made with specific providers in the country (as has often been the case in countries where permits are not required for R&D, or where there are no ABS regulations), including local communities.

What seems clear regarding national permit aspects of ABS systems is that they need to achieve a balance between monitoring and controlling the activity of researchers, while also allowing for the conduct of research – particularly where it contributes to the conservation and sustainable use of biodiversity. As Parry (2004) noted, many institutions and governments in developing countries are now developing 'detailed legislative and administrative structures that, they are told, will enable them to secure ongoing returns from bioprospecting operations. While there is now probably a greater degree of realism among these governments and less discussion of 'green gold' in the relevant international forums, this is an important point. She also added her concern that this process (which may well fail) may direct finite resources (human capital, infrastructure and money) away from other crucially important areas like health and education (Parry, 2004) and other direct conservation actions.

Enabling versus protective national systems

A number of commentators have noted that there are ABS systems that can be considered 'enabling' versus others that might be considered 'protective/ defensive.' For example, the permitting and ABS systems of Brazil and the Philippines have been described in papers and in the halls and side-events of the CBD COP as being defensive to the point where they are substantially restricting research activities on biodiversity in those countries (e.g. Dávalos et al., 2003; Filoche, 2012). Other countries have sought to reach a middle ground or to be enabling for researchers, whether they have non-commercial or commercial intent: South Africa and Australia can probably be described in this way. For the great majority of countries there is not a clear system of ABS in place – this lack of legal certainty for both users and providers was a key driver for the development of the Nagoya Protocol.

The Nagoya Protocol essentially encourages a 2-track system. Article 9 on 'special considerations' asks Parties to 'create conditions to promote and encourage research which contributes to the conservation and sustainable use of biological diversity, particularly in developing countries, including through simplified measures on access for non-commercial research purposes ...' This inclusion reflects agreements typical of the ICBG. The article has been emphasised because of regular concern and complaint from biological researchers who are often subjected to the same scrutiny as bioprospectors, but who have purely scientific intent that generates knowledge which is often beneficial for conservation. The Article also notes the need to consider change of intent – national permitting needs to clearly define research allowed and the process that researchers must follow if they intend to switch to commercial research. This would normally necessitate a new agreement with MAT and a different array of benefit-sharing (often milestones and royalty payments). Change of intent for biotrade transfers raise a slightly different issue (discussed below).

Ultimately the concerns and policy of the country towards bioprospecting should frame the way they draft their ABS laws. Countries are within their 'sovereign rights' to establish restrictive regimes, should they choose to. This is likely to often shift activities to neighbouring countries or regions, where there may be more receptive policies. Indeed, it seems likely that some states will allow for an 'open access' policy (either explicitly or by default of not creating access requirements). This seems likely to be the case in the European Union which has just adopted ABS regulations in 2014 that do not define rules for access to genetic resources in member states. The EU regulations leave it open to member states as to whether or not to require PIC and benefit-sharing for genetic resources in their territories. This may be because much of Europe has relatively low endemic species diversity and because its countries are more often user countries. If EU member states decide not to develop ABS rules for access, this may be attractive to researchers and companies because it is less 'red tape' and bureaucracy. On the other hand, some companies and researchers will be hoping to conduct research where there is legal certainty regarding their actions, and clearly specified rules for access and PIC should provide this.

The potential for open access policies has been criticised by some, regarding the implications for genetic resources in collections and genebanks that have been acquired by Europe-based institutions. The EU regulations impose no retroactivity on genetic resources and associated TK acquired from other countries before the entry into force of the EU ABS regulations. This is despite many developing country negotiators seeking to include specific text on 'temporal scope' that would include certain attempts for retroactive benefit-sharing, at least to the time that the CBD came into force. Since the 'green revolution' much agricultural genetic resources were collected from the biodiverse developing countries of the world and have been acquired by various private and public collections in Europe, the United States and in other foreign genebanks. If there are no regulations surrounding the transfer of materials and extracts from these past collections, then arguably past injustices may continue to be perpetuated. Individual institutions, particularly public institutions, typically make material transfer agreements that specify a number of terms of use. However, without sufficient agreement about intended use, benefit-sharing, monitoring and compliance mechanisms, the movement, use and third party use of genetic resources obtained prior to enactment of these regulations, it will be hard to enforce these aspects, if they will be enforced at all. For example, Biber-Klemm et al. (2014) note the Conservatoire et Jardin Botaniques de la Ville de Geneve, in Switzerland, receives an average of 15,000 specimens per year and exchanges on loan about 8,000 specimens with more than 100 institutions worldwide. Herbarium specimens are typically exchanged for purely academic or taxonomic research, but there has been concern that on occasions, commercial research is conducted (as discussed in Chapter 2; see Parry, 2004, pp. 175–199). The scale of international transfers for herbarium specimens alone highlights the impracticalities of trying to monitor everything that is transferred, reiterating the importance of *intent for R&D*, particularly for commercial purposes, should be specified in permits and transfer agreements. There would then be reliance on the ability to check the validity of permits in the ABS Clearing House (and this raises questions about the functionality of the still-pending Clearing House to achieve this). It also highlights that attempts to enact retro-active benefit-sharing would become extremely complex and impractical and so it is not surprising that they have not been pursued in the

European regulations. Aside from concerns about retroactivity, there is some concern about the 'presumption of legality' of the 'register of collections of the Union' (i.e. trusted collections), given that it only applies for users obtaining genetic resources for 'utilisation' (the term 'use of genetic resources' means 'utilisation for R&D' – Article 3). This means that others could conceivably obtain samples for commercial use or as part of the herbarium trade and change intent – what happens if this then occurs outside the European Union (see ICTSD, 18 March 2014) (discussed below in compliance)? There is an assumption that the EU regulations will have a normative effect on other countries and partners with whom genetic resources are exchanged. The due diligence requirements for users to ascertain that the genetic resources and associated TK has been accessed in accordance with applicable legal requirements is intended to cover concerns about compliance (Article 4). However, it is not clear exactly when the due diligence requirement would end – it specifies 20 years after the 'end of the period of use' (Article 4.3). Does this mean after a product reaches commercialisation and is no longer 'utilised for R&D'? Why 20 years?

It is not only the European Union that has a relatively open access policy. De facto open access has been the case in some states in Australia (the sub-tropical states), despite some original intent for a 'nationally consistent' system for ABS. In these states, ABS seems to have failed to gain sufficient political attention – which is remarkable, and perhaps also foolish, given that Australia has high endemic biodiversity. The fact that the northern states are located in tropical climates and cover somewhat 'less developed' regions may reflect their greater interest in capitalising on the utilisation of genetic resources. This may also be in part due to higher concern about TK appropriation in the Northern Territory, Queensland and Western Australia which have relatively higher concentrations of indigenous populations and more communities living relatively traditional lifestyles.

Widely held traditional knowledge

In an insightful paper, Ruiz Muller (2013) noted that transboundary TK is the prevailing rule, rather than the exception, in the context of indigenous peoples' cultures and livelihoods. Yet the focus of much policy-making and research, including at times my own (see Robinson and Kuanpoth, 2009), has tended to focus on the idea that TK is held within certain discrete boundaries and communities. In the context of this book, we are interested in biodiversity-related TK for medicinal, food or other uses (as differentiated from cultural expressions or folklore). However, this differentiation is something that is often quite artificial and has in many ways been furthered by the way intellectual property separates knowledge, information and expressions into different forms. Indigenous groups might highlight that TK of certain plants is transferred through folklore and stories, for example. Ruiz Muller cites a recent bracketed and working definition of TK from the WIPO Intergovernmental Committee (IGC) draft articles for TK protection:

know-how, skills, innovations, practices, teachings and learning of [indigenous [peoples] and [local communities]]/[or a state or states] that are dynamic and evolving, and that are intergenerational/and that are passed on from generation to generation, and which may subsist in codified, oral or other forms.

(Ruiz Muller, 2014)

This working definition highlights the commonly evolving and changing nature of TK, and that it is passed between individuals and groups. Inevitably, with these transfers, the knowledge is conveyed into neighbouring communities, to neighbouring countries or even regions or continents. While it is inherently difficult to generalise about TK, there is some inevitability that it could be documented and transferred out of oral forms and into the written public domain for use by others.

In the context of ABS, Ruiz Muller (2013 , 2014) raises 3 challenges affecting the ability to negotiate an agreement when TK is widely distributed or shared:

1   How can a contract be negotiated, or PIC obtained, when there is no single, clearly defined right holder?
2   Even if this were possible, assuming TK is in the public domain or publicly available, is it feasible and economically viable to negotiate an advantageous contract?
3   What are the effects of economic pressures on benefit potential, when TK is in practice accessible from various sources?

These sorts of dilemmas have plagued certain ICBG projects, including the Maya and Peru projects mentioned in Chapter 11, including the related question of who has 'established rights to provide access to genetic resources'. The Samoan mamala case is relevant here too. Although Dr Paul Cox obtained specific knowledge of the medicinal uses of the plants from individual healers, a number of commentators in Samoa have noted that this knowledge was probably held by many people in Samoa, and potentially in other countries in Polynesia. Yet he established agreements with the community at Falealupo (first), and with the government and individuals (at a later date), noting specific individuals as beneficiaries of royalties and milestone payments should they arise. In this way, the benefit-sharing might be established with the indigenous and local community providers of the knowledge, even if it is widely held. The risk here is that there may be competing claims over the origins of the knowledge and the benefit-sharing solely with 'providers' may not be seen as fair and equitable by other TK holders or groups. The same can be raised of the Moroccan argan case in which the agreements were established with the EIG Targanine group and Prof. Charrouf, despite there being widely held TK of the uses of argan oil in cosmetics. In this case, it is important to note that the benefits accruing to the social fund of

Targanine's cooperatives relate to a new piece of R&D on the by-product of argan pressing, meaning that TK may have only played an indirect and partial role in this lead.

The paper by Ruiz Muller (2014) suggests a few different protection options. The first is trade secrets, which has been suggested in the past and has some advantages (it is a widely recognised form of IP protection used in many jurisdictions) and disadvantages – it is costly to register, registration doesn't guarantee prevention of disclosure (especially where the knowledge is widely held by various groups), it is thus better suited to narrowly held TK, it is a formal protection that may be beyond the understanding and affordability of remote or very 'traditional' communities, and it may lead to the exclusion of communities or individuals who hold similar knowledge but are not involved in the given contracts. He poses a second option that might be useful for knowledge already in the public domain – the copyright derived principle of 'domaine public payant' (Muller, 2014). This applies to widely shared works, crafts and arts that have lost copyright protection but are deemed important enough to receive special policy attention obliging a fee for their use. Furthermore, the Nagoya Protocol (Article 12) and a number of NGOs such as Natural Justice and the Compass Network have encouraged the development of (bio-cultural) community protocols. However, these are more relevant for specific communities, and it establishes the community's conditions by which their genetic resources, TK, and potentially other resources or assets (e.g. land, water, sites and artworks) might be used. The limitations of community protocols are that they generally do not have binding legal effect; however, they may establish a pathway and lay out clear expectations that then lead to contracts.

Ruiz Muller (2013) also pointed to the use of Article 10 of the Nagoya Protocol, which specifies the intent to develop a global multilateral benefit-sharing mechanism. This might be somewhat like the mechanism the ITPGRFA has for its multilateral system, but which relates to genetic resources for food and agriculture – to date this has barely been used to receive income from benefit-sharing; instead it has been disbursing grants derived from country donations. An international compensatory fund of this sort has previously been suggested by authors such as Drahos (2000), but he suggested it as a global bio-collecting society, where knowledge would be registered and those bioprospecting or seeking knowledge could contact the society. They would then be put into contact with relevant groups for negotiation of an agreement – where this idea becomes problematic is again where the knowledge is very widely held, rather than just several communities. This idea has the advantage that it is established without the involvement of states and thus brokers collector to knowledge-holder group agreements – something desirable for many indigenous groups who feel left out of many agreements and who may have antagonistic relationships with the state. Ruiz Muller (2013) and many others (Kuanpoth and Robinson, 2009; Rangnekar, 2009; and in Blakeney et al., 2012) have suggested the use of geographical indications protection. This is, however, limited to the protections of specific goods or products that might be produced in a specific region and that have special qualities (often due to specific know how, traditions, practices and knowledge as well as growing conditions in that region). It may also only be useful and cost-effective for specific well-established industries such as Darjeeling Tea or Souss valley Moroccan argan. Other approaches have also been suggested, but as Dutfield (2003, 2013) noted, unless these are flexible enough to encompass the diversity of customary laws and practices relating to access to and use of TK, any new international norms will likely fail.

Secret, sacred and locally held traditional knowledge

As Ruiz Muller (2014) notes, certain TK is 'still secretly guarded by designated leaders in communities – figures such as the shaman the elder, or the healer'. However, sometimes comparable knowledge might be found in neighbouring communities or villages that have similar ecosystems and biodiversity (and often shared or inter-twined histories). There are certain circumstances whereby TK may be held by relatively few individuals, either deliberately or because of the lack of inter-generational transfer of this knowledge and associated practices (which is an all-too-common story in local villages and communities throughout the Pacific and Southeast Asia, if not many regions of the world). As individuals and communities become less 'traditional' and stop certain practices and rituals, they quickly begin to lose their knowledge associated with the use of biological resources. With the many advances in information and communications technology, different entertainment options and distractions from 'traditional life' in even very remote locations are opening up. For example, during fieldwork in Chiang Mai in Northern Thailand, I have spent time with Karen and Hmong communities which reside in quite remote forest areas and who primarily live subsistence lifestyles – farming rice and vegetables, breeding livestock and hunting for most of their daily consumption needs. However, it is not unusual to see the people from these villages talking on mobile phones, the children wearing Manchester United shirts or Chelsea Football Club shirts and trying to hitch a ride into town on the back of a motorbike to watch TV, play soccer or buy sweets (see Box 11.1). The idea of the 'ecologically noble savage' upon which much discussion of TK is still based, is not only outdated, it also poorly understands that indigenous and local community identities and lifestyles are very often cosmopolitan and blend the traditional and modern. In the ABS context, this is relevant for at least 2 reasons: there are researchers interested in gaining access to local communities and documenting TK (including knowledge related to biological resources) before it is lost; there might be ethnobotanists or bioprospectors interested in identifying useful knowledge and/or trying to identify appropriate beneficiaries to negotiate with (e.g. see the Samoa case where specific individual knowledge holders were written into benefit-sharing agreements).

Box 11.1 Traditional knowledge in the context of modern life in Karen villages in Chiang Mai

In 2006 and 2009, I conducted fieldwork with my research assistant 'Ae' in some Karen and Hmong villages in Chiang Mai. The intention was to understand how local TK of biodiversity was locally regulated and controlled, and if customary laws and rules existed. One of the villages – Baan Soplan in Samoeng district was particularly interesting.

Upon entering the village, it is evident that the community is heavily reliant upon their surrounding environment for subsistence. The people live in raised, bamboo and timber houses, cut from the forest, and they are largely set into the forest under the canopy. Nearby are shifting rotational agriculture fields for rice and vegetable production. Surrounding the houses of the village is some livestock – mainly chickens and pigs. The people also hunt boar and other animals in the surrounding forest, and forage for foods and medicines. While some of the elders are in basic traditional hand-sewn clothes, most people are in t-shirts and shorts or wrap-skirts. Noticeably the young adults, teenagers and children are wearing Manchester United shirts or Chelsea Football Club shirts. While the village looks quite 'traditional', it is not unusual to see the people from these villages talking on mobile (cell) phones – there is one functional solar panel installed by an NGO and it is used mainly for charging people's mobile phones and for watching the one TV in the village. Although most people live and work in the forest, some do travel into town for work (typically manual labour but also the sale of small goods like medicines, fabrics or handicrafts) either by hitching a ride into town or on the back of a motorbike. Some of the children also hitch rides to high school, or just to watch TV, go to market, play soccer or buy sweets in town.

From my interviews, it was evident that the elders in the community had considerable knowledge of plants and medicines from the surrounding forest. They explained the use of several medicines and the customary laws surrounding their use (different aspects are discussed in Robinson, 2010 at pp. 110–111; Robinson and Kuanpoth, 2009; and in considerable detail in Robinson, 2012). The elders explained that knowledge is often shared and passed on to friends and family. People come when they are sick and need treatment, and that is when it is usually shared. However, there are various customary laws which guide who they share it with, when and for what purposes. These relate to animist spiritual beliefs (and occasionally to Buddhist beliefs). Notably, a number of these medicines should not be traded or treated as a commodity. This knowledge builds by doing some experimentation – but there are rules for a reason and some people in the past have died through their experiments. Some knowledge was very specific to Karen communities in the region. Other knowledge of medicinal plants was derived from adjacent regions of Thailand – for example knowledge of the *Kwau Krua* vine reputedly came to the village from *Issan* or the Northeast. This particular vine is in the public domain, has been documented and used for decades however, and even certain uses are patented. But other plants are known and used locally and kept relatively secret according to customary laws (Patthi Ta Yae and Patthi Daeng, 2–3/7/2009).

The elders worried about the loss of this knowledge. Patthi Ta Yae would like to see a school established where people could be taught about traditional medicines from the forest, and to stimulate interest among the youth. He notes that they are generally disinterested in the traditional way of living and a number of them had moved into nearby towns. Some elders had more detailed knowledge than him, but have since passed away (e.g. his grandfather). New healers are usually chosen according to a dream or a ritual. Some people who have these dreams ignore them because they do not want to be tied down to the many rules that accompany being a village healer, and the training involved (the learning of holy words, knowledge of plants, recipes, treatments and symptoms). This knowledge is normally kept all in their head – it is not written down. This has been deliberate because if holy words are written incorrectly, the treatment will not work (Patthi Ta Yae, 2/7/2009). However, Patthi Daeng noted that it would be useful to write the holy words and the correct treatments so that they are not lost and forgotten (Patthi Daeng, 3/7/2009).

As discussed in Robinson (2008) and Forsyth (2012, 2013), protection to prevent biopiracy and/or encourage fair and equitable benefit-sharing does not necessarily *promote* TK. As knowledge is lost between generations a number of other approaches might be required including secret or open registers or databases, fairs and events that celebrate this knowledge, incentives for local industries based on this knowledge and through the greater assertion of local rights and customary laws that allow the continuation of traditional practices.

A number of approaches that might be relevant are discussed in the previous section and in Robinson (2010). Another example that is worth mentioning and which is relevant to specific, secret and sacred TK is the Cook Islands Traditional Knowledge Policy and Bill. The purpose of the policy is to:

Ensure mechanisms are in place to protect, preserve and promote the Cook Islands Traditional Knowledge nationally, in the Pacific region, and internationally. This policy recognizes that traditional knowledge belongs to the indigenous communities of the Cook Islands whose ties to the 15 islands can be traced back traditionally and over many generations. This policy guides further work in relation to Traditional Knowledge by signaling key principles and objectives. This policy is a first step towards Government acknowledging its social responsibility to work in partnership with its indigenous communities to protect, preserve and promote this knowledge for the sake of current and future generations.

(Cook Islands Herald, 20/4/2011)

Thus, the Cook Islands government is clarifying that TK belongs to indigenous communities, relevant to the implementation of aspects of the Nagoya Protocol detailed in Articles 7 and 12. Furthermore, the policy and subsequent bill appear to be registering traditional knowledge so that it is retained and not lost, and so that it can also be promoted and protected. The policy appears to have been influenced by concerns about misuse, unauthorised reproductions and/ or biopiracy in the region and further explains:

> This policy is the first step to putting past selfish practices in the past and ensures the process of benefit sharing is made abundantly clear so the Owners benefit from the use, practice and care of their knowledge . . . Today, traditional knowledge is held by indigenous communities and used by many as part of daily living practices. Some of this knowledge has been compromised by modern inventions, skills and creations recreating the knowledge and experience of the users of this knowledge. Nonetheless traditional knowledge is now recognized as an important and valuable resource for indigenous people internationally and efforts are being made to protect, preserve and promote it.

(Cook Islands Herald, 20/4/2011)

In response to the policy a Traditional Knowledge Bill was drafted in 2013. We had the opportunity to meet with the staff of the Ministry of Culture in November 2013 who will implement the Act once passed by Parliament. The Bill has been designed as a sui generis form of national protection, intended to also link to agreements made at the regional level (there are regional model laws of TK and cultural expressions that were developed by the Pacific Islands Forum Secretariat, with the assistance of WIPO). The definition of traditional knowledge provided is:

. . . either of the following:

(a) knowledge (whether manifested in tangible or intangible form) that is, or is or was intended by its creator to be, transmitted from generation to generation; and—

   i   originates from a traditional community; or
   ii is or was created, developed, acquired, or inspired for traditional purposes:

(b) any way in which traditional knowledge appears or is manifested ...

This definition then includes a long list of manifestations in songs, rituals, designs, processes and other forms, meaning that they are using the term in the broadest possible sense (i.e. it includes the WIPO-synonymous categories of traditional knowledge, cultural expressions and folklore). The act would apply retrospectively to TK existing before it comes into force, but does not affect contracts or IP rights relating to TK existing before the act. The law encourages the registration of TK. The law would assert rights of the TK to the creators of the knowledge or their successors, including rights of transmission, use, commercialisation, acknowledgement (essentially as moral rights) and others. The law then restricts non-rights holders against similar actions in relation to registered knowledge (e.g performing or exhibiting the work in public without permission). Registered TK is protected in perpetuity, the rights are inalienable and purport not to limit or affect other intellectual property rights (although conceivably there are circumstances where issues and conflicts might arise regarding the assertion of IP rights).

Interestingly, the law prohibits certain actions relating to registered secret-sacred knowledge, including use, transmission, receiving commercial benefit for use or development, without written authorisation of the rights-holder. However, the question arises – how would someone know if they have done this if the knowledge is secret? The staff at the Ministry of Cultural Development indicated that the researcher can check with key staff involved in maintaining the registry.

Also relevant to the discussion in the last section, the law anticipates multiple and overlapping registrations. If 2 or more registrations for ostensibly the same knowledge are made, then each applicant is able to view the others application and must come to an agreement before registration is accepted for protection under the law.

This law seems an attempt to respond mainly to the protection of knowledge held by specific groups and individuals. Despite some aspects encouraging the potential commercial development of products from TK, the policy and law have defensive protection implications. This leads to a question about 'freedom to operate' for businesses or communities who sell certain products based on the knowledge, especially where there are multiple and overlapping claims or where the knowledge is widely held.

During discussions with these staff and other Cook Islanders, it was noted that there is often a considerable difference between the way people live on the main island of Rarotonga (quite developed with many large tourist facilities, many foreign residents or visitors and a port and township – Avarua), and most of the other islands, particularly those that are most remote (living more traditional lives with semi-subsistence livelihoods). This law thus appears to be seeking a balance between protecting and promoting TK of those who might be most vulnerable to its loss, and those who seek to commercialise products based on TK, practices and designs. As mentioned earlier, there is one ABS agreement in the Cook Islands, between a Cook Islands researcher and an indigenous representative body – the Koutu Nui. In discussions with the researcher – Dr Graham Matheson – he emphasised that Cook Islanders, including the Koutu Nui, are often very business savvy, and that it should not be assumed that they are naïve, even if they do retain TK and practices (see Robinson, 2013).

Thus the Cook Islands are clearly a TK provider country, but are also a user country. On this note, it is worth noting one aspect of the EU regulations – largely a user region, though not without its own TK (often well-documented and public domain). There has been some criticism of the lack of reference to PIC of indigenous and local communities where access is sought to genetic resources (see Article 6.2 of the Nagoya Protocol) and associated TK (see Articles 7 and 12 of the Nagoya Protocol). As discussed in Chapter 2, there has been some consternation and concern from indigenous and local communities that the wordings of these articles in the Nagoya Protocol are too weak because of phrases such as 'in accordance with domestic law' and 'as appropriate'. The EU regulations have a fairly ambiguous piece of text in Article 4, which indicates:

> Users shall exercise due diligence to ascertain that genetic resources and traditional knowledge associated with genetic resources used were accessed in accordance with applicable access and benefit-sharing legislation or regulatory requirements and that, where relevant, benefits are fairly and equitably shared upon mutually agreed terms. Users shall seek, keep, and transfer to subsequent users information relevant for access and benefit-sharing.
>
> (EU Draft Regulations on ABS, Article 4)

This indicates that the European Union have taken a narrow view of the Nagoya Protocol articles which describe the rights of indigenous peoples and local communities, with the regulations obliging that their users only have to comply with ABS regulations on PIC and MAT. The issue here will be with how quickly countries can establish ABS regulations and if indeed they will include specific regulations which recognise the rights of indigenous peoples and local communities over genetic resources and associated TK, and require local PIC and MAT. This is particularly relevant where we are concerned about secret, sacred or locally held TK. We might also question the likelihood of there being specific local PIC requirements embedded into regulations in countries such as China or Russia, for example, which have questionable human rights records in relation to minority groups.

As discussed in Chapter 2, the inclusion of elements for the potential recognition of indigenous rights over genetic resources, associated TK and customary law are all significant achievements for indigenous empowerment. However, the weak language leaves open the potential for continued denial or dilution of indigenous rights in many circumstances, leaving this a matter for ongoing struggle at the local and national level.

Mutually agreed terms, power relations and 'fairness and equity'

Having discussed TK of indigenous and local communities, it is worth also turning to the idea of 'MAT' briefly, and considering the issue of power relations. When seeking to establish a contract that specifies the terms of utilisation, intellectual property rights, benefit-sharing and dispute settlement among other things, the idea of MAT has been encouraged since the CBD was first developed, so as to ensure 'fairness and equity'. However, some have been critical about how likely this is when you are often talking about disparities of power. Agreements are often made with governments who control access to genetic resources, but also often made at a local level with providers. The examples in this book show a range of providers with arguably quite different capabilities to negotiate. There have been governments, government agencies or national institutes (Madagascar, Thailand, Samoa, Australia, Vanuatu, PNG and Panama) and also local communities (Madagascar, Morocco, Samoa, Vanuatu and PNG) involved in making agreements about ABS. While we assume that governments usually have the appropriate legal expertise and knowledge of ABS to make an *informed* decision, this may not always be the case, as there are often limited capacities in the departments of environment (or similar) that deal with ABS – just in terms of numbers of staff, knowledge of ABS, level of education, familiarity with contracts. There may also be political bias, corruption and related issues that influence a decision. For example, some anonymous interviewees in Samoa were critical of the agreements made by the government because of the perception of bias.

Regarding local communities, these might also have limited capacity and knowledge of ABS, affecting their ability to negotiate a fair and equitable agreement. Gradually this should improve as countries implement the Nagoya Protocol and start informing communities about their role and rights. Projects of the Global Environment Facility and its Small Grants Programme have begun targeting this concern by encouraging awareness raising. Also, there are programmes such as the multi-donor GIZ-led ABS Capacity Development Initiative that work in the African, Caribbean and Pacific regions to create awareness of ABS and build local capacities, often with the support of the CBD Secretariat and other bodies such as the ITPGRFA. Even further, NGOs such as Natural Justice have been assisting communities with the development of community protocols to specifically frame community desires, concerns, customary norms and laws and rights with respect to ABS. These sorts of initiatives plus the programs of governments should improve local ABS negotiation capacity over time, and national laws will or should often specify that local informed consent allow sufficient time for local providers to obtain legal advice (as is required in some of the Australian ABS laws). This improves the likelihood of greater 'fairness and equity' but does not guarantee that everyone will be happy. In most circumstances there is a strong likelihood that non-beneficiaries will feel left out. For example, in the Samoan mamala case and the Moroccan argan case, there was some evidence from interviews of envy or concern about fairness and equity. A review of the Samoan press over several years suggests that tensions between and within communities and across Samoa have simmered about the agreements.

Access and upfront benefit-sharing

The language of the Nagoya Protocol is for 'benefits arising from the utilization of genetic resources as well as subsequent applications and commercialization shall be shared in a fair and equitable way with the party providing such resources ...' The term 'arising from' suggests that sharing of benefits occurs *after* the 'utilisation' or R&D. This might typically mean milestone payments at key stages of development and commercialisation, as well as royalty payments. However, this is not strictly the case, and should probably have been written as 'relating to' the utilisation of genetic resources (and associated traditional knowledge). To explain, the Annex of the Nagoya Protocol contains a list of several monetary and non-monetary benefits that might be shared upfront. These include 'access fee/fee per sample collected', 'up-front payments', 'research funding', 'special fees to be paid to trust funds supporting conservation or sustainable use of biodiversity', 'collaboration ... in scientific research and development', 'human and material resources to strengthen capacities for the administration and enforcement of access regulations', 'training relating to genetic resources' and others. Thus, ABS might also contribute some benefits at the point of access or shortly afterwards. For many who are cynical about ABS, it is because agreements rarely deliver upfront monetary benefits, and instead speculate that they might arise 15–20 years down the track. More commonly, there is upfront training, technology transfer and small-scale benefits in the short term. Certainly, for academic projects – including large expeditions like the Santo 2006 project – there is a very limited budget to be spent on core research activities and relatively little to provide to local communities otherwise. For these projects, the non-monetary benefits upfront are crucial to generating local goodwill about a project. It is particularly useful if the project can have some positive benefits for education or conservation as the Santo project has tried to accomplish.

If we consider the Madagascan ICBG and the Moroccan argan case, these both have significant short-term monetary benefits for communities. In both of these cases the indigenous and local communities involved generally expressed very high levels of satisfaction in interviews with us (albeit a limited sample set for which adjacent non-beneficiaries might not be so happy). These projects were also exceptional from the cases examined, because they were able to achieve this level of upfront benefit through considerable industry involvement and investment (the Novartis cases instead focused on upfront training and technology transfer). While some authors are critical of ABS (discussed in Chapter 1) specifically because it can be perceived as a way for industry to privatise and commodify nature by reducing it down to genes or biochemical extracts, there is probably a greater likelihood of an ABS agreement satisfying local concerns and desires if it is designed broadly to encompass these local needs and local conservation (like the current ICBGs are typically designed) and it has sufficient investment to fund this 'broadness' (like the Madagascar and Moroccan argan cases). In other words, it is harder to be cynical about bioprospecting projects with ABS agreements in place, when the 'users' are spending hundreds of thousands of dollars in the local communities for purchases and projects that the community has decided on, as is the case in both the ICBG Madagascar and Moroccan argan cases.

Utilisation and benefit-sharing

This leads us to consider the ongoing role of milestone payments and royalties. While it seems unlikely that that will disappear from ABS agreements, particularly for pharmaceutical drug discovery which has the longest R&D pipeline, the experiences in the cases in this book (and others like the Hoodia case) suggest a need to de-emphasise their role. Many governments, authors and academics are now working on template and model agreements for ABS. This is the time to be more creative with contracts and to think carefully about the expectations of provider countries and communities. Where agreements are made with genebanks and collections these are likely to continue to follow a milestone/royalty payment approach that may provide benefits in the long term given that there are low chances of success.

However, other industries such as nutraceuticals, cosmetics and related skin/health care have different concerns about image. They typically continue to source raw products from nature after R&D has been conducted and products become commercial, meaning that they may also have consumers scrutinising this sourcing. At conferences of the Union for Ethical Biotrade (UEBT), this potential overlap between benefit-sharing for utilisation and fair and ethical natural product supply arrangements is often discussed. The Moroccan argan case provides an interesting example of how these might be handled in tandem, and how benefits towards a community fund or social fund might be funded (by paying far higher than market price for the product that has been utilised for R&D and which is used to continue to derive extracts from), value-added (technology transfer for basic processing and quality control is conducted at Targanine) and distributed (to an in-country cooperative economic interest group – EIG – which is run to benefit the local women 'shareholders'). While this has worked for use in a high-end skin-care product sold by one of the world's largest companies in this sector, it may not be so easy for companies with smaller budgets for R&D and investment, smaller margins and for lower cost products. However, something similar could be attempted at a smaller scale.

Monitoring and checkpoints

The designation of checkpoints for monitoring was tensely negotiated towards the Nagoya Protocol, with many wanting the inclusion of patents specifically mentioned. However, the text remains quite open as to what Parties to the CBD might use. Patents are one of the most logical places for an ABS checkpoint because they are a signifier of R&D (to be eligible, the inventor must demonstrate an inventive step, typically requiring some R&D), and because patents are exclusive rights of monopoly that have been the main point of consternation since the CBD came into force (or even earlier than that). As the cases in Robinson (2010) demonstrate, misappropriations of genetic resources have caused some concern in the past, but much more significant public reactions and protests in developing countries and among indigenous, local and farmer populations have occurred because of the use of intellectual property rights, typically patents. Patents represent the ability to restrict others from using a genetic resource for a specific purpose, and thus invoke a potentially harmful economic restriction on freedom to operate for 'providers'. Notably, this has only rarely occurred, but still the potential for this to happen has generated significant discontent. Furthermore, as I explain in Robinson (2010; see also Robinson et al., 2014), it is often the political and cultural impact of a patent over a genetic resource and associated TK that generates the ire and labels of 'biopiracy'. This cultural dimension has been poorly understood by many Northern researchers and arrogantly dismissed (see Heald, 2003; and particularly Chen, 2006).

Thus a requirement for disclosure of source of origin, or proof of legal provenance of genetic resources for registration of a patent, has been pursued by many countries. The WTO TRIPS council has now received several submissions by countries towards an amendment of TRIPS to include a disclosure requirement. A detailed proposal was put forward in 2008 on 'draft modalities on TRIPS-related issues' that included a proposal to move forward with an amendment of TRIPS to include a disclosure requirement, from Albania, Brazil, China, Colombia, Ecuador, the European Communities, Iceland, India, Indonesia, the Kyrgyz Republic, Liechtenstein, the Former Yugoslav Republic of Macedonia, Pakistan, Peru, Sri Lanka, Switzerland, Thailand, Turkey, the ACP Group and the African Group, plus Croatia, Georgia and Moldova. A later submission was made in 2011 by a group of developing countries for an amendment to TRIPS Article 29*bis* to include specific text on disclosure. However, despite some buy-in from developed countries such as those in Europe, there is still considerable disagreement in TRIPS and the WTO in general and negotiations have stagnated since this time. Indeed, several countries around the world have various versions of a patent disclosure requirement. For European countries, there has been concern surrounding the mandatory nature of a disclosure requirement and the penalty for non-compliance. Developing countries have sought the rejection of a patent for non-compliance. These proposals have not been well-received by the United States and Japan, who have argued that it would unnecessarily burden the patent system, which is already strained, causing further delays to examinations. This, they argue is unfair to industry, who are seeking to protect their innovations, and these 2 countries have stated a preference for a contract-based approach for resolving ABS transactions. However, while contracts can and will be used for ABS agreements, provisions for monitoring in an ABS contract may be meaningless unless there are better ways for interested members of the public to check how and where genetic resources and associated genetic resources have been used.

There are other checkpoints that can be used, such as university ethical clearances, grant-making bodies and publication houses or journals. In fact, these all already do monitor various aspects of researcher behaviour. However, these mechanisms are not likely to be effective for monitoring research by companies, because they rarely publish their findings (they keep them secret for competitive advantage), and they do not typically rely on grants but instead seek private investment.

Compliance

The EU Draft Regulations specify the monitoring of user compliance through reporting on public research funding, where there are requests for marketing approval for a product developed from genetic resources or traditional knowledge, or at the time of commercialisation (to be reported to the competent national authority). Competent authorities then transmit this information to the ABS Clearing House every 2 years. This seems a clear attempt to avoid mandating use of the patent system to monitor compliance to EU member states, at least until further consensus in the WTO TRIPS Council or WIPO can be obtained. The competent authority is then allowed to perform checks to ensure users are in compliance with due diligence obligations and instances where declarations are obliged to be made, including where concerns are substantiated by third parties. They can then also check for internationally recognised certificates of compliance at the Clearing House for evidence of PIC and MAT as required by the ABS regulations or policy of another country. The penalties that they can impose are fines, suspension of specific use activities and confiscation of illegally acquired genetic resources.

While the Clearing House will likely create a significant additional measure of transparency for checking user compliance, there will still need to be a vigilant public to notify national authorities of the user and provider countries to draw attention to potential non-compliances. However, it might still conceivably be difficult for a member of the public, or even a trained lawyer, to identify exactly when an organisation might have conducted R&D on a genetic resource or associated TK, based solely on marketing information. Marketing information is typically superficial, and while it may include ingredient lists, for say cosmetics or herbal medicines, it may not for pharmaceuticals (how are we to know if specific biochemical compounds come from plants or are synthetic?). Even for cosmetics, herbals and nutraceuticals, it might be difficult to know if R&D has been conducted. In fact, some from these industries have queried the threshold for 'R&D' because many companies might examine the biochemical composition of a genetic resource, but then include it as a raw entity, or a refined extract into a product. This is another reason why a patent disclosure of origin requirement makes a useful compliance monitoring juncture –because it normally contains much of this information required to prove innovation for the grant of the patent.

Intellectual property

One gaping absence from the Nagoya Protocol is any substantive reference to intellectual property. The negotiators from several developing countries sought to have patents acknowledged as a checkpoint (aligning with their requests for a disclosure of origin requirement in TRIPS); however, this was ultimately negotiated out of the final text amidst a bunch of compromises. Even reference to the role of WIPO was watered down in Article 4, which contains general text about the relationship of the Nagoya Protocol with other International agreements and instruments.

Patents and other forms of intellectual property have often been suggested as forms of protection for genetic resources and associated TK; however, there are obvious limitations on when these might be used. Indeed, Oldham et al (2013) questioned the contribution that patents had made to the exploration of biodiversity and suggested the need to think more broadly about incentive measures. For many the costs of filing a patent are prohibitive and there may be little benefit as a result, unless seeking to gain some commercial advantage offered by the exclusive right. This means that they are less likely (though not completely improbable) to be used for the defensive protection of genetic resources or associated TK. For example, in Thailand some universities have filed for patents over certain uses of genetic resources as a defensive strategy against potential foreign uses, while allowing Thai users free access to the technology. In other cases, researchers that have utilised certain genetic resources and associated TK have acknowledged the indigenous contributors as co-inventors and/or joint patent holders (e.g. the Griffith University and Jarlmadangah Buru people and their agreement over an analgesic; the Cook Islands CIMTECH – Koutu Nui case). But, while some indigenous groups and local communities may accept these forms of protection, there are others who would not. For example, in Robinson (2010) I discuss the Hawaii Taro case in which indigenous Hawaiians campaigned against a patent on taro on cultural and spiritual grounds. When the researchers agreed to hand over ownership of the patent to the indigenous people concerned, they refused, noting the significance of the plant to them and that no patent should be held over taro (see also Robinson et al., 2014).

While some other forms of intellectual property have been suggested for the protection of TK, such as trade secrets and geographical indications, the evidence to date is that these expensive Western modes of protection have not been heavily utilised by indigenous and local communities. Rather, the cases of geographical indications that do offer protection over TK, are often for products from industries set up by colonial powers – such as the Darjeeling and Ceylon Tea cases, and in the Moroccan argan case (Berber or Amazigh TK that has been capitalised on by Arab and European industries, but also appropriated by Israeli companies). Thus, while these might form part of a potential suite of protection options, they are unlikely to be commonly used for genetic resource and traditional knowledge protection, unless there is an already established industry from which benefits are likely.

In parallel to the work being conducted to develop and implement the Nagoya Protocol, there has been considerable work on the development of draft texts in the WIPO IGC on Intellectual Property and Genetic Resources, Traditional Knowledge and Folklore. For several years, these texts have been driven forward with the intent of developing international instruments for the protection of genetic resources, traditional knowledge, and cultural expressions and folklore as separate instruments. The negotiations have progressed significantly, despite difficult questions such as regarding definitions, duration of protection, on disclosure of origin, on prevention of misappropriation, if nations as well as communities would be beneficiaries, amongst many other things (New, 2011; Ghosh, 2012). The exclusion of indigenous people from direct participation in negotiations of direct relevance to their knowledge has been a bitter issue for their representatives. Despite a fund which supports travel expenses for indigenous people to attend, the negotiations are still state-based, and the indigenous groups are recognised as observers (New, 2011; Vivas-Eugui, 2012). This has led to a number of angry statements from indigenous representatives about their concerns being ignored and texts being changed without their consultation and despite their objections. In 2012, the nature of the instruments sparked much debate as to whether they would be legally binding (sought by the Like Minded Group of Developing Countries) or flexible and non-binding (the EU) or yet to be determined because of the need for further refinement and removal of bracketed text (the Group of Central European and Baltic States) (ICTSD, 2012).

Others have suggested *sui generis* approaches to protection. These often ultimately mirror Western/Northern modes of protection. For example, the Cook Islands Traditional Knowledge Bill is a unique approach, but it also shares some traits with forms of copyright protection – and was influenced somewhat by WIPO technical assistance (see Ghosh, 2012 for a detailed discussion). Similarly the Pacific Regional Model Laws on Traditional Knowledge received assistance from WIPO, and were drafted by lawyers from the Pacific after considerable consultations with Pacific countries, but to date have had little uptake. I suspect that the problem with such sui generis systems, no matter how well-designed, may be that they appear to indigenous and local as external impositions and that they often duplicate or would overlie existing customary laws and norms that regulate knowledge transfer and use of nature. As put by David Vivas-Eugui (2012, p. 25), talking of the WIPO texts:

> Most indigenous and local communities have not really been demanding a sui generis system to protect GRs in their territories, their TK and TCEs, but rather a clear recognition of a wider set of rights, including self-determination, human rights, customary law, and land rights. Also, policies for ensuring the preservation of TK and their livelihoods are high on their agenda. Benefit-sharing arising from the utilization of TK and TCEs only comes after these first two priorities.
>
> (European Commission, 2014, Article 4)

The lack of basic recognitions and rights needs to be overcome as a first step in the overall process towards traditional knowledge protection. This is one reason why people have been working with communities to develop community protocols from 'the ground up', such that the community's own regulations and expectations are conveyed to external parties (who might otherwise be completely ignorant of local rules). The steps towards formal recognition of customary protocols and laws are also critical for indigenous communities.

References

Biber-Klemm, S., Davis, K., Gautier, L. and Martinez, S.I (2014) 'Governance options for *ex-situ* collections in Academic Research'. In Oberthür, S. and Rosendal, G.K. (eds) *Global Governance of Genetic Resources: Access and Benefit-Sharing after the Nagoya Protocol.* Routledge, Oxford, pp. 213–230.

Blakeney, M.L., Coulet, T., Mengistie, G. and Mahop, M. (2012) *Extending the Protection of Geographical Indications: Case Studies of Agricultural Products in Africa.* Routledge, London.

Chen, J. (2006) 'There is no such thing as biopiracy... and it's a good thing too'. *McGeorge Law Review,* 37(1), 1–32.

Chiarolla, C., Lapeyre, R. and Pirard, R. (2013) *'Bioprospecting under the Nagoya Protocol: A Conservation Booster?'* Available at: www.iddri.org, accessed 28/4/2014.

Cook Islands Herald. *'Traditional Knowledge to be Protected'.* Available at: http://www.ciherald.co.ck/articles/h560d.htm, accessed 16/4/2014.

Dávalos, L.M., Sears, R.R., Raygorodetsky, G., Simmons, B.L., Cross, H., Grant, T., Barnes, T., Putzel, L. and Porzecanski, A.L. (2003). 'Regulating access to genetic resources under the Convention on Biological Diversity: An analysis of selected case studies'. *Biodiversity & Conservation,* 12(7), 1511–1524.

Drahos, P. (2000) 'Indigenous knowledge, intellectual property and biopiracy: Is a global bio-collecting society the answer?' *European Intellectual Property Review,* 22(6), 245–250.

Dutfield, G. (2003) *Intellectual Property and the Life Science Industries: A Twentieth Century History.* Ashgate, Aldershot, UK.

Dutfield, G. (2013) 'Transboundary resources, consent and customary law'. *Law, Environment and Development Journal,* 9(2), 259–263.

European Commission. (2014) Regulation (EU No 511/2014), 16 April 2014, on compliance measures for users from the Nagoya Protocol on Access to Genetic Resources and the Fair and Equitable Sharing of Benefits Arising from their Utilization in the Union Text with EEA Relevance.

Filoche, G. (2012) 'Biodiversity fetishism and biotechnology promises in Brazil: From policy contradictions to legal adjustments'. *The Journal of World Intellectual Property,* 15(2), 133–154.

Forsyth, M. (2012) 'Do you want it giftwrapped? Protecting traditional knowledge in the Pacific Island countries'. In Drahos, P. and Frankel, S. (eds) *Indigenous Peoples' Innovation: Intellectual Property Pathways to Development.* ANU ePress, Canberra, pp. 189–214.

Forsyth, M. (2013) 'How can traditional knowledge best be regulated? Comparing a proprietary rights approach with a regulatory toolbox approach'. *The Contemporary Pacific,* 25(1), 1–31.

Gamez, R. (2007) 'The link between biodiversity and sustainable development: Lessons from INBio's bioprospecting programme in Costa Rica'. In McManis, C. (ed) *Biodiversity and the Law: Intellectual Property, Biotechnology and Traditional Knowledge*. Earthscan, London, pp. 77–90.

Ghosh, S. (2012) 'The quest for effective traditional knowledge protection: Some reflections on WIPO's recent IGC discussions'. *Bridges Trade BioRes Review*, 6(2). Available at: http://ictsd.org/i/news/bioresreview/135678/, accessed 28/4/2014.

Heald, P.J. (2003) 'The rhetoric of biopiracy'. *Cardozo J. Int'l & Comp. Law*, 11, 519–546.

ICTSD. (2012) 'WIPO assemblies approve IGC roadmap'. *Bridges Trade BioRes Review*, 6(4), 15 October. Available at: http://www.ictsd.org/bridges-news/biores/news/wipo-assemblies-approve-igc-roadmap, accessed 15/7/2014.

ICTSD. (2014) 'WIPO genetic resources talks advance, though differences linger'. *Bridges Trade BioRes Review*, 18(5), 13 February, 13. Available at: http://www.ictsd.org/bridges-news/bridges/news/wipo-genetic-resources-talks-advance-though-differences-linger, accessed 15/7/2014.

Kuanpoth, J. and Robinson, D. (2009) 'Geographical indications protection: The case of Thailand and Jasmine Rice'. *Intellectual Property Quarterly*, 2009(3), 288–310.

New, W. (2011) 'New text in play in WIPO traditional knowledge, genetic resources talks'. Available at: www.ip-watch.org/, accessed 27/03/2014.

Oldham, P., Hall, S. and Forero, O. (2013) 'Biological diversity in the patent system'. *PLOS One*, 8(11), 1–16.

Parry, B. (2004). *Trading the Genome: Investigating the Commodification of Bio-information*. Columbia University Press, Columbia.

Rangnekar, D. (2009) 'Indications of geographical origin in Asia: Legal and policy issues to resolve'. In Melendez-Ortiz, R. and Roffe, P. (eds) *Intellectual Property and Sustainable Development: Development Agendas in a Changing World*. Edward Elgar, London, pp. 273–302.

Robinson, D. (2008) 'Beyond "protection": Promoting traditional knowledge systems in Thailand'. In Gibson, J. (ed) *Patenting Lives: Life Patents, Development and Culture*. Ashgate, Aldershot, UK, pp. 121–138.

Robinson, D. and Kuanpoth, J. (2009) 'The traditional medicines predicament: A case study of Thailand'. *Journal of World Intellectual Property*, 11(5/6), 375–403.

Robinson, D. (2010) *Confronting Biopiracy: Cases, Challenges and International Debates*. Routledge/Earthscan, London.

Robinson (2012) 'Towards Access and Benefit-Sharing Best Practice: Pacific Case Studies.' DSEWPAC, Canberra and GIZ, Eschborn, Germany. Available at: http://www.abs-initiative.info/uploads/media/ABS_Best_Practice_Pacific_Case_Studies_Final_01.pdf, accessed 15/7/2014.

Robinson, D.F. (2013) 'Legal geographies of intellectual property, "traditional" knowledge and biodiversity: Experiencing conventions, laws, customary law, and karma in Thailand.' *Geographical Research*, 51, 375–386.

Robinson, D.F., Drozdzewski, D. and Kiddell, L. (2014) ' "You can't change our ancestors without our permission": Cultural perspectives on biopiracy'. In Fredrikksson, M. and Arvanitakis, J. (eds) *Piracy – Leakages of Modernity*. Litwin Books, Sacramento, CA, pp. 55–75.

Rosenberg, B. (2006) 'Market concentration of the transnational pharmaceutical industry and the generic industries: Trends on mergers, acquisitions and other transactions'. In Roffe, P., Tansey, G. and Vivas-Eugui, D. (eds) *Negotiating Health: Intellectual Property and Access to Medicines*, Earthscan, London, pp. 65–78.

Ruiz Muller, M. (2013) 'Protecting shared traditional knowledge: Issues, challenges and options'. *ICTSD*, Issue Paper 39, Geneva. Available at: http://ictsd.org/i/publications/179919/, accessed 28/04/2014.

Ruiz Muller, M. (2014) 'Protecting widely shared traditional knowledge'. *Bridges Trade BioRes Review,* 8(2). Available at: http://ictsd.org/i/news/bioresreview/185629/, accessed 28/04/2014.

Tobin, B. (2013) 'Bridging the Nagoya compliance gap: The fundamental role of customary law in protection of indigenous peoples' resource and knowledge rights' *Law. Environment and Development Journal,* 9(2), 142–162.

Vivas-Eugui, D. (2012) 'Bridging the gap on intellectual property and genetic resources in WIPO's intergovernmental committee (IGC)'. *ICTSD,* Issue Paper no 34, Geneva.

12 Conclusions

The real test for 'fair and equitable' benefit sharing from bioprospecting will be closely related to the adoption of the Nagoya Protocol by certain key actors, and will depend *how* they adopt it. The 'elephant in the room' is the United States which is unlikely to ratify the CBD and adopt the Nagoya Protocol. Given the size of its domestic biotechnology, pharmaceutical, health care, cosmetic and other industries and the location of many multinationals companies there, it is likely that there will continue to be a huge gap in adoption of ABS principles and practice and user compliance. While there are likely to be some, like the ICBG-funded projects, who will continue to do in-situ bioprospecting in foreign countries, will seek prior informed consent, establish mutually agreed terms with local providers and monitor the utilisation of genetic resources and extracts that are collected. There are also likely to be others in the United States who will still be able to avoid (naively, deliberately or both) ABS negotiations with providers and governments in other countries, or that will simply use domestic genebanks and collections to source genetic resources. Even if US researchers continue to pursue in-situ biodiscovery activities in countries that have Nagoya compliant ABS regulations, if they do not comply there may be little legal action that can be taken except for the cancellation and prevention of future permits. The worst punishment might be 'naming and shaming', but only if the provider country, concerned members of the public or NGOs are able to identify that a non-compliance has occurred. The Clearing House mechanism may assist with this; however, there are still relatively few ways to check on non-compliance (usually by scrutinising the filing of a patent or publication of research findings) and these can be laborious. Even if checked, patent documents and publications do not always disclose how or where genetic resources are obtained, or if traditional knowledge is used.

Another main set of actors – the countries located in the European Union – have begun putting in place regulations for the adoption of the Nagoya Protocol. However, we have noted earlier that there are limits on the scope of the EU regulations in the form of 'due diligence' obligations on users, no retroactivity for samples already in the European Union, and an emphasis on 'trusted collections'. This is likely to have the effect of encouraging re-examination of existing genetic resources already in genebanks. I anticipate that this sort of regulatory change will also further shift utilisation of traditional knowledge towards public domain sources and avoidance of retroactivity. This means that in-situ foreign country bioprospecting by EU-based researchers and companies will likely decline even further, as predicted by Parry (2004) (e.g. Swiss-based Novartis already has been targeting foreign collections that seem tightly regulated and probably fit the concept of 'trusted collections'). This is despite evidence from patent analysis that current innovative activity focuses on only about 4 per cent of taxonomically described species and between 0.8 and 1 per cent of predicted global species (Oldham et al., 2013) – suggesting there is wide scope for further natural products-based innovation, including further in-situ collection. If we judge by the cases in this book, further re-mining of existing collections also probably means a focus on training and technical non-monetary benefits with only the small likelihood of long-term monetary royalty or milestone benefits (at least for pharmaceuticals). Further genebank sourcing might have the effect of reducing payment of upfront benefits, besides small access fees, to local communities in foreign countries (often with direct conservation application) that have been discussed in some of the in-situ biodiscovery examples in this book (e.g. the ICBGs discussed, the Samoan case, the INBio cases). The exception might be through the cosmetic industry and like industries who perceive a marketing advantage in ethical and sustainable sourcing, including fair and equitable benefit sharing. Other areas that might continue doing field bioprospecting could include crop protection and agrochemicals (e.g. Dow is currently supporting ICBG projects), and ornamental horticulture for botanicals which still has some incentive to search for wild varieties for breeding, though this is probably quite low (CBD Secretariat, 2008).

The trends in innovation utilising biological diversity noted by Oldham et al. (2013) and the general trend towards sourcing from ex situ collections are worrying for several reasons. This is likely to reduce the flow of direct benefits to local communities because much germplasm and genetic resources have been collected in the past. This will complicate the benefit sharing process through to original providers and see genebanks as the main recipient of benefits – though these are often public institutions that also have an important conservation role in their own right. This trend away from in situ bioprospecting and thus direct benefit-sharing unfortunately goes against one of the original reasons for the ABS conditions – to seek to redress a colonial imbalance of 'Northern extraction' and 'Southern oppression' that goes back decades to the green revolution period which had medicinal parallels (see Osseo-Asare, 2014) and back centuries to the spice trade (see Robinson, 2010, chapter 1; Schiebinger, 2004; and Parry, 2004, chapter 2). We might see the further entrenchment of the devolution of R&D onto public institutions like universities which take on the risk and challenges of bioprospecting, despite often being relatively poorly funded, with companies only coming in to purchase or licence research at a later date. If adequate 2-stage contracts are not established with providers, then little benefit may trickle back to them. It might also mean further 'patent evergreening' and development of 'me-too' drugs (often for everyday conditions and lifestyle diseases rather than tropical or neglected diseases) by companies utilising intellectual property rights to cling to any and every advantage, rather than taking risks that might be of greater benefit to society globally (see Kuanpoth, 2006; Rosenberg, 2006; Widdus, 2006).

Indeed, some might blame the intellectual property system and those that have sought to enhance it and globalise it for its inability to adapt to certain changing conditions and impacts, despite its global reach and effects, and despite the extension of patent terms and other commercial advantages. Indeed many countries, such as the African Group and Bolivia, have called for the repeal of patentability of genetic resources outright, on moral and cultural grounds. Others have pointed out that the scope of patentability of genetic materials and chemical isolations has expanded through successive judgements, and that even if companies are unable to secure a patent, that filing is a priority in its own right and can be a significant commercial deterrent (as mentioned in Chapter 4). Indeed, there are a number of excellent histories written about the misuse of the patent system towards the establishment of cartels, the consolidation of global industries (e.g. pharmaceutical, agricultural and chemical) and indeed the deterrence of innovation by all but the largest few (see Palombi, 2009; Braithwaite and Drahos, 2000; Drahos with Braithwaite, 2003; Dutfield, 2003). It seems that, at the very least, there is some need to reform patent law such as through a disclosure requirement (Oldham et al., 2013), or to dramatically overhaul the system in an era that is supposed to be about free trade but which is reaching new levels of protectionism (Palombi, 2009).

A useful point is made by Ruiz Muller (2013) that much effort has gone into the establishment of an ABS regime to protect specifically held genetic resources and traditional knowledge, when much of it is widely held, publicly accessible or distributed broadly. This does not mean that the Nagoya Protocol is a failure, but rather that those implementing it need to be realistic about what can be captured, controlled and what benefits might accrue. For others, the Nagoya Protocol is more about the incremental assertion of rights. On a final and more personal note, I have worked with indigenous, local and farming communities for about 10 years now. From my own work on traditional knowledge, I agree with authors such as Posey et al. (1996), Dutfield (2013) and Tobin (2013) that political space needs to be ceded such that indigenous people's assertions are heard and they have the rights they deserve to basic things like land, or resources or political representation. It is hard to imagine that communities can protect their knowledge and the genetic resources that they use, if they do not have adequate recognition of their rights to land or sea areas. Indeed, states will find it hard to protect genetic resources supposedly under sovereign control if they have not made efforts to involve indigenous and local communities in the design of their protections, including through the formal recognition of customary protocols and laws. For indigenous populations, the implementation of the Nagoya Protocol has the potential to be one step in a larger empowering process – let us hope so.

References

Braithwaite, J. and Drahos, P. (2000) *Global Business Regulation.* Cambridge University Press, Cambridge.

CBD Secretariat (Laird, S. and Wynberg, R.). (2008) *Access and Benefit-Sharing in Practice: Trends in Partnerships Across Sectors.* CBD Secretariat, Montreal, Technical Series No. 38.

Drahos, P. with Braithwaite, J. (2003) *Information Feudalism: Who Owns the Knowledge Economy?* New Press, New York.

Dutfield, G. (2003) *Intellectual Property and the Life Science Industries: A Twentieth Century History.* Ashgate, Aldershot.

Dutfield, G. (2013) 'Transboundary resources, consent and customary law'. *Law, Environment and Development Journal,* 9(2), 259–263.

Kuanpoth, J. (2006) 'TRIPS-Plus intellectual property rules: Impact on Thailand's public health'. *The Journal of World Intellectual Property,* 9(5), 573–591.

Oldham, P., Hall, S. and Forero, O. (2013) 'Biological diversity in the patent system'. *PLOS One,* 8(11), 1–16.

Osseo-Asare, A.D. (2014) *Bitter Roots: The Search for Healing Plants in Africa.* University of Chicago Press, Chicago.

Palombi, L. (2009) *Gene Cartels: Biotech Patents in the Age of Free Trade.* Edward Elgar, Cheltenham.

Parry, B. (2004). *Trading the Genome: Investigating the Commodification of bio-information.* Columbia University Press, Columbia.

Posey, D.A., Dutfield, G., Plenderleith, K., da Costa e Silva, E. and Argumedo, A. (1996) *Traditional Resource Rights: International Instruments for Protection and Compensation for Indigenous Peoples and Local Communities.* International Union for the Conservation of Nature (IUCN), Gland.

Robinson, D. (2010) *Confronting Biopiracy: Cases, Challenges and International Debates.* Routledge/Earthscan, London.

Rosenberg, B. (2006) 'Market concentration of the transnational pharmaceutical industry and the generic industries: Trends on mergers, acquisitions and other transactions'. In Roffe, P., Tansey, G. and Vivas-Eugui, D. (eds) *Negotiating Health: Intellectual Property and Access to Medicines.* Earthscan, London, pp. 65–78.

Ruiz Muller, M. (2013) *Protecting Shared Traditional Knowledge: Issues, Challenges and Options,* Issue Paper 39. ICTSD, Geneva. Available at: http://ictsd.org/i/publications/179919/, accessed 28/4/2014.

Schiebinger, L. (2004) *Plants and Empire: Colonial Bioprospecting in the Atlantic World.* Harvard University Press, Cambridge.

Tobin, B. (2013) 'Bridging the Nagoya compliance gap: The fundamental role of customary law in protection of indigenous peoples' resource and knowledge rights'. *Law, Environment and Development Journal,* 9(2), 142–162.

Widdus, R. (2006) 'Product development partnerships on "neglected diseases": Intellectual property and improving access to pharmaceuticals for HIV/AIDS, tuberculosis and malaria'. In Roffe, P., Tansey, G. and Vivas-Eugui, D. (eds) *Negotiating Health: Intellectual Property and Access to Medicines.* Earthscan, London, pp. 205–226.

Appendix 1
List of interviews and Personal Communications

The list is represented by chapter and in chronological order.

Chapter 3: Madagascar

Naritiana Rakotoniaina (ABS focal point), interview, Service d'Appui a la Gestion de l'Environnemont (SAGE), Antananarivo, 02/09/2013.

Chris Birkinshaw (MBG), interview, Missouri Botanical Garden (MBG), Antananarivo, 02/09/2013 and 06/09/2013.

Dr Luciano Andriamaro (ABS program director for CI), interview, Conservation International, Antananarivo, 02/09/2013.

Professor Felicitee Rejo, interview, Madagascan Centre Nationale de Recherches sur l'Environment (CNRE), Antananarivo, 03/09/2013.

Dr Rado Rasolomampianina, interview, CNRE, Antananarivo, 03/09/2013.

Dr Michel Ratsimbason, interview, Madagascan Centre Nationale d'Application et des Recherches Pharmaceutiques (CNARP), 04/09/2013.

Mamisoa Andrianjafy (MBG), interview, Ibity and surrounds, 05/09/2013.

Ten men at Sahamalola, interviews, Ibity and surrounds, 05/09/2013.

Four men and two women at Sahanivotry, interviews, Ibity and surrounds, 05/09/2013.

Three women and seven men at Manandona, interviews, Ibity and surrounds, 05/09/2013.

Six men at Ibity town/commune, interviews, 05/09/2013.

Jeannie Raharimampionona (MBG), interview, Ankazobe and surrounds, 06/09/2013.

The head of the Fokontany, the head of the parents and citizens association, the teacher, and about 25 parents of children at the school at Firaranzana, focus group-style interview, Ankazobe and surrounds, 06/09/2013.

Six men from the local Firaranzana forest conservation group, interview, Ankazobe and surrounds, 06/09/2013.

Jeremie Razafitsalama (MBG), interviews, Antsiranana and Oronjia peninsula, 07/09/2013 and 09/09/2013.

Seven fishermen at Ankorikihely, interviews, Oronjia peninsula, 07/09/2013.

Five men at Ambodivahibe, interviews, Oronjia peninsula, 07/09/2013.

Seven women and two men at Ambavarano, interviews, Oronjia peninsula, 09/09/2013.

Two women at Ambodimanga, interviews, Oronjia peninsula, 09/09/2013.

Prof David Kingston, Pers. Comm. (email), 22/10/2013 and 20/01/2014.

Chapter 4: Thailand

Dr Kanyawim Kirtikara (BIOTEC), interview, Bangkok, 19/06/2013.
Sronkanok Tangjaijit (BIOTEC), interview, Bangkok, 19/06/2013.
Dr Bubpha Techapattaraporn (BIOTEC), interview, Bangkok, 19/06/2013.
Mr. Koichi Yoshino (Shiseido), interview, Bangkok, 19/06/2013.
Several NGOs and academic commentators, pers. comm., Bangkok, 20-21/06/2013.
Frank Petersen (Novartis), pers. comm. (email), 17/09/2013.
NOTE: other BIOTEC and Shiseido staff also attended the interview.

Chapter 5: Samoa

Solia Papu Va'ai, Apia, 13/03/2012.
Meeting with several Falealupo Matai: Fuiono Aleki, Taii Tapana, Tapua Tamasi, Manutuaifo, Kelemete, Gaga Sanele, Ulufanua Aleuna, pers. comm, Falealupo, Savaii, 15/03/12.
Kolone Va'ai, Matai, interview, Falealupo and Vaisala, Savaii, 15/03/2012.
Fuiono Patolo, Matai, interview, Falealupo, Savaii, 23/05/12.
Seumantufa Fale mai, Matai and healer, interview, Falealupo, Savaii, 23/05/12.
Anonymous interview, Samoa, Falealupo, Savaii, 23/05/12.
Lemau Seumantufa, elder and healer, interview, Falealupo, Savaii, 24/05/2012.
Manu Toifotino, Matai, interview, Falealupo, Savaii, 240/5/2012.
Taii Tulai, Matai, interview, Falealupo, Savaii, 24/05/2012.
Marianive Fuiono, Elder from Falealupo, Falealupo, Savaii, 24/05/12.
Aeau P. Leavai, Member of Parliament for Falealupo (current and past), Apia, 25/05/12.
Tima Leavai, Lawyer from Falealupo, Apia, 25/05/12.
Anonymous meeting notes and comments by Samoan contributors, Second Pacific ABS Workshop, ABS Initiative, Asau, Savaii, 22-25/05/2012.
Paul Cox. pers. comm. (email) 06/06/2012.
Clark Peteru, pers. comm., various dates, 2012-2013.
NOTE: a further three interviews were undertaken and provided context towards this case study. The participants have not been named for ethical reasons.

Chapter 6: Morocco

Rachel Barre (L'Oreal), pers. comm., Paris, 26/07/2010.
Charlotte D'Erceville (BASF), pers. comm., Paris, 26/07/2010.
Professor Zoubida Charrouf (University of Rabat), interviews, Agadir and Taitmatine - Taroudant, 03-04/08/2010.
Interview with the manager and assistant manager of the Taitmatine Cooperative, Taroudant, 04/08/2010.
Focus group-style interview with several women of the Taitmatine Cooperative, Taroudant, 04/08/2010.
Interviews and meetings with Targanine staff including Latifa Anaouche (Business Manager of Targanine), Agadir and surrounds, 26, 28 and 29/04/2011.
Interview with the manager, assistant manager, and technical staff of the Tagmate Cooperative, Imouzzar prefecture, 25/04/2011.

Focus group-style interview with several women of the Tagmate Cooperative and associated centres, Imouzzar prefecture, 25/04/2011.

Interview with the manager and assistant manager of the Tamaynoute Cooperative, Agadir Idaoutanane, 26/04/2011.

Interview with the manager and informal discussions with several women of the Targante Cooperative, Chtouka Ait Baha Province, 27/04/2011.

Interview with the manager, accountant and three women of the Ajdigue Cooperative, and three additional women at an associated centre, Province de Chtouka, 28/04/2011.

Interview with the Manager and assistant manager of the cooperative Toudarte, Imsouane region, 29/04/2011.

Focus group-style interview with several women of the cooperative Toudarte, Imsouane region, 29/04/2011.

NOTE: this chapter was also informed by personal communications with Eric D'efrenne (Yamana), and Raquel Ark (formerly at Cognis).

Chapter 7: Australia

Geoff Burton (UNU and IES, UNSW), pers. comm. Several dates from 2012-2014.

Richard Lipscombe (Proteomics International), pers. comm. (email) 24/06/2011.

NOTE: This chapter also benefited from several presentations and the discussion at the DSEWPAC Oceania Biodiscovery Forum, held at the Eskitis Institute, Griffith University on 19-21 November 2012.

Chapter 8: Vanuatu

Dr Philippe Bouchet, French National Museum of Natural History, Paris, pers. comm. (Email). 04/06/12.

Donna Kalfatak, Department of Environmental Protection (Biodiversity Unit), interview, Port Vila, 11/06/12.

Norman, J. Head Librarian, Vanuatu Cultural Centre, pers. comm., Port Vila, 11/06/12.

Hickey, F. Vanuatu Cultural Centre (phone), pers. comm., Port Vila, 11/06/12.

Anonymous, former research/field assistant for the Santo project, pers. comm., 12/06/12.

Rufino Pineda, Former National Coordinator for the Santo Expedition, interview, Port Vila, 12/06/12.

Russell Nari, Former Director General of the Department of Lands, interview, Port Vila, 12/06/12.

Joel Path, Secretary General of Sanma Province (Santo), interview, Luganville, 13/06/12.

Obeid, A. Director of Fisheries Office in Santo, pers. comm., Luganville, 13/06/12.

Serei Maliu, Chief of Butmas Village, plus several other men, interview, Santo, 14/06/12.

Chief Solomon, Chief of Matantas/Vatthe, interview, Santo, 14/06/12.

Purity Solomon, interview, Matantas/Vatthe, Santo, 14/06/12.

Bill Tavine, interview, Matantas/Vatthe, Santo, 14/06/12.

Chapter 9: Papua New Guinea

Professor Louis Barrows, University of Utah, pers.comm. (Email), 15/12/2011, 06/06/12, 08/06/12 and numerous other dates from 2011-2013.

Professor Lohi Matainaho, University of Papua New Guinea, pers.comm. (Email), 15/12/2011, (presentation and discussion) Asau, Savaii, Samoa, 23/5/2012, and numerous other dates from 2011-2013.

Dr Eric Kwa, Director of PNG Law Reform Commission, pers. comm.(presentation and discussion) Asau, Savaii, Samoa, 23/05/2012, and 20/11/12.

Chapter 10: Panama

Professor Phyllis Coley, University of Utah, pers. comm., (email) 06/04/2014.
Professor Thomas Kursar, University of Utah, pers. comm., (email) 06/04/2014.
Dr Joshua Rosenthal, FIC/ICBG programmes, pers. comm., (email) 27/03/2014.
Dr Flora Katz, FIC/ICBG programmes, pers. comm., (email) 01/04/2014.

Index

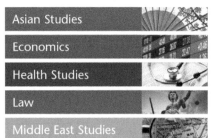